VLSI Electronics Microstructure Science

Volume 1

Contributors

J. M. Ballantyne

D. K. Ferry

Jay H. Harris

Patrick E. Haggerty

R. W. Keyes

Michael C. King

M. P. Lepselter

W. T. Lynch

Bruce R. Mayo

Charles H. Phipps

Fred W. Voltmer

E. D. Wolf

VLSI Electronics Microstructure Science

Volume 1

Edited by

Norman G. Einspruch

School of Engineering and Architecture
University of Miami
Coral Gables, Florida

1981

ACADEMIC PRESS

ACADEMIC PRESS

A Subsidiary of Harcourt Brace Jovanovich, Publishers

New York   London   Toronto   Sydney   San Francisco

6572-6038 ✓

CHEMISTRY

ACADEMIC PRESS, INC.
111 Fifth Avenue, New York, New York 10003

United Kingdom Edition published by
ACADEMIC PRESS, INC. (LONDON) LTD.
24/28 Oval Road, London NW1 7DX

Library of Congress Cataloging in Publication Data
Main entry under title:

VLSI electronics: Microstructure science.

    Includes bibliographical references and index.
    1. Integrated circuits--Large scale integration.
I. Einspruch, Norman G.
TK7874.V56   621.381'73   81-2877
ISBN 0-12-234101-5 (v. 1)   AACR2

PRINTED IN THE UNITED STATES OF AMERICA

81 82 83 84   9 8 7 6 5 4 3 2 1

To
EDITH
with love

Contents

Chapter 1   Manufacturing Process Technology for MOS VLSI
Fred W. Voltmer

Chapter 2   Principles of Optical Lithography
Michael C. King

* Section II.A by George Gamota; Section II.B by John O. Dimmock; Section III.A by W. Murray Bullis and Robert I. Scace; Section III.C by James A. Hutchby; Section IV.A by Myles G. Boylan; Section IV.B by William A. Hetzner; Section IV.C by Stanley Pogrow.

List of Contributors

Numbers in parentheses indicate the pages on which the authors' contributions begin.

J. M. Ballantyne (129), National Research and Resource Facility for Submicron Structures, Cornell University, Ithaca, New York 14853

Myles G. Boylan (265), Division of Policy Research and Analysis, National Science Foundation, Washington, D.C. 20550

W. Murray Bullis (265), Electron Devices Division, National Bureau of Standards, Washington, D.C. 20234

John O. Dimmock (265), Physics Division, Office of Naval Research, Washington, D.C. 20301

D. K. Ferry (231), Department of Electrical Engineering, Colorado State University, Fort Collins, Colorado 80523

George Gamota (265), Office of the Undersecretary of Defense for Research and Engineering, Department of Defense, Washington, D.C. 20301

Patrick E. Haggerty* (301), Texas Instruments Incorporated, Dallas, Texas 75265

Jay H. Harris† (265), Division of Electrical, Computer, and Systems Engineering, National Science Foundation, Washington, D.C. 20550

William A. Hetzner (265), Division of Policy Research and Analysis, National Science Foundation, Washington, D.C. 20550

James A. Hutchby (265), Energy and Environmental Research Division, Research Triangle Institute, Washington, D.C.

R. W. Keyes (185), Thomas J. Watson Research Center, International Business Machines Corporation, Yorktown Heights, New York 10598

Michael C. King (41), Microlithography Division, The Perkin-Elmer Corporation, Norwalk, Connecticut 06856

* Deceased.

† Present address: College of Engineering, San Diego State University, San Diego, California 92182.

M. P. Lepselter (83), Bell Telephone Laboratories, Murray Hill, New Jersey 07974

W. T. Lynch (83), Bell Telephone Laboratories, Murray Hill, New Jersey 07974

Bruce R. Mayo (301), Texas Instruments Incorporated, Dallas, Texas 75265

Charles H. Phipps (301), Texas Instruments Incorporated, Dallas, Texas 75265

Stanley Pogrow (265), Division of Policy Research and Analysis, National Science Foundation, Washington, D.C. 20550

Robert I. Scace (265), Electron Devices Division, National Bureau of Standards, Washington, D.C. 20234

Fred W. Voltmer (1), Intel Corporation, Santa Clara, California 95051

E. D. Wolf (129), National Research and Resource Facility for Submicron Structures, Cornell University, Ithaca, New York 14853

Preface

Civilization has passed the threshold of the second industrial revolution. The first industrial revolution, which was based upon the steam engine, enabled man to multiply his physical capability to do work. The second industrial revolution, which is based upon semiconductor electronics, is enabling man to multiply his intellectual capabilities. VLSI (Very Large Scale Integration) electronics, the most advanced state of semiconductor electronics, represents a remarkable application of scientific knowledge to the requirements of technology. This treatise is published in recognition of the need for a comprehensive exposition that describes the state of this science and technology and that assesses trends for the future of VLSI electronics and the scientific base that supports its development.

These volumes are addressed to scientists and engineers who wish to become familiar with this rapidly developing field, basic researchers interested in the physics and chemistry of materials and processes, device designers concerned with the fundamental character of and limitations to device performance, systems architects who will be charged with tying VLSI circuits together, and engineers concerned with utilization of VLSI circuits in specific areas of application.

This treatise includes subjects that range from microscopic aspects of materials behavior and device performance—through the technologies that are incorporated in the fabrication of VLSI circuits—to the comprehension of VLSI in systems applications.

The volumes are organized as a coherent series of stand-alone chapters, each prepared by a recognized authority. The chapters are written so that specific topics of interest can be read and digested without regard to chapters that appear elsewhere in the sequence.

There is a general concern that the base of science that underlies integrated circuit technology has been depleted to a considerable extent and is in need of revitalization; this issue is addressed in the National

Research Council (National Academy of Sciences/National Academy of Engineering) report entitled "Microstructure Science, Engineering and Technology." It is hoped that this treatise will provide background and stimulus for further work on the physics and chemistry of structures that have dimensions that lie in the submicrometer domain and the use of these structures in serving the needs of humankind.

I wish to acknowledge the able assistance provided by my secretary, Mrs. Lola Goldberg, throughout this project, and the contribution of Academic Press in preparing the index.

Contents of Other Volumes

Chapter **1**

Manufacturing Process Technology for MOS VLSI

FRED W. VOLTMER

Intel Corporation
Santa Clara, California

I. INTRODUCTION

The requirements for VLSI manufacturing process technology are defined by the circuit design rules inherent in the implementation of VLSI functions and by the continued decrease in the cost per function. In this chapter, the manufacturing process technology required to implement MOS VLSI designs in production will be described. Emphasis will be placed on the manufacturing aspects of the processes, and research and development will largely be ignored. Although VLSI bipolar integrated circuits are not discussed, many of the primary findings about MOS VLSI can be applied to bipolar circuits also.

The manufacturing of integrated circuits includes die fabrication and die assembly. The entire semiconductor manufacturing process flow is illustrated in Fig. 1. To manufacture VLSI integrated circuits, changes will occur predominantly in the fabrication of the semiconductor die, not in the assembly of the die into a package unit, although some changes will be required in assembly to accommodate the larger die. Thus, only the manufacturing process technology of fabricating the die will be discussed in this chapter. The generally accepted definition for VLSI of 100K gates, or memory bits per circuit, will be used. The chapter will begin with a review of the trends in circuit density, since these trends lead to the definition of the process technology. Methods of achieving the circuit density and performance will be related to both traditional horizontal scaling and emerging vertical scaling. The resultant pervasiveness of low-temperature and dry plasma processing will be summarized.

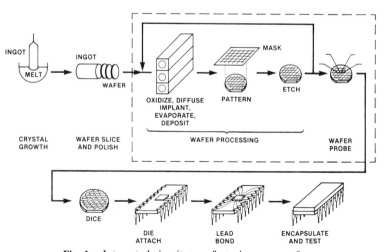

Fig. 1. Integrated circuit manufacturing process flow.

In Section II attention will be given to historical trends, which give insight into the manufacturing requirements for VLSI. The subsequent sections will review the technology and manufacturing implications of these trends. The focus of Sections III and IV will be on scaling and its implications for manufacturing. It is through scaling that VLSI can and will be realized. Section III examines traditional horizontal scaling and includes a review of lithographic techniques and dry etching techniques. Extension of the capability of each of these is essential to the realization of VLSI. The technological details of lithographic techniques are covered independently in Chapters 2 and 3 of this volume, while plasma processing is covered in Volume 2, Chapter 1.

Section IV addresses vertical scaling, an area that has until recently been given little attention. In this section, the requirements for achieving high-quality thin oxides, shallow junctions, and low-resistivity interconnects will be reviewed.

In Section V the requirements for manufacturing process control and in-process characterization will be discussed. Section VI assesses trends in wafer size and the implications of larger-diameter wafers as well as silicon material quality requirements and changes in photomask requirements.

The considerable new capacity added for the manufacture of VLSI circuits and the trends in the nature of this capacity will be described in Section VII. Manufacturing support system requirements and the status of automation will also be examined.

The intent of this chapter is to outline the trends in technology that must be comprehended by the manufacturer in order to produce VLSI circuits, and an attempt is made to show the relationship among the various changes taking place in manufacturing technology requirements. However, the technological details required to manufacture VLSI circuits will not be provided.

II. TRENDS IN COMPLEXITY

The factors contributing to the ever-increasing circuit density of MOS integrated circuits leading to VLSI have been described by Moore [1] and examined in more detail by Keyes [2]; the physical limitations to continued circuit density growth have been discussed by Hoeneisen and Mead [3]. In Fig. 2, the trends in complexity outlined by Moore are shown, along with the following factors contributing to the increased complexity: (1) smaller feature size, (2) increased die size, and (3) increased design cleverness. Increases in die size and reduction in feature

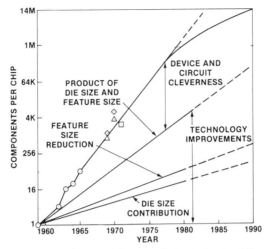

Fig. 2. Contributions to increased integrated circuit density (after Moore [1]).

size provide for the major portion of the increase in components per chip. Thus the challenge of the manufacturer is to comprehend the manufacturing implication of Moore's law and to provide the capability to fabricate, in high volume, circuits with ever-increasing die size, ever-decreasing feature size, and ever-increasing complexity. In evolving toward VLSI, it is anticipated that these trends will continue.

Consider the implication of each of these components contributing to increased density as they relate to the manufacture of integrated circuits. To manufacture VLSI circuits cost-effectively, a properly designed circuit and adequately developed process are required. When a circuit and process are properly designed and the circuit parameters fall within the process tolerances, then defects become the limiting item in the ability to manufacture the integrated circuit. This defect-limited yield of integrated circuits has been shown to be a strong function of die size [4–6]. Thus, as the die size increases to implement VLSI, the defect density of the manufacturing facility must be lowered.

Feature size, historically, has referred to the smaller horizontal dimension, and the reduction of feature size has centered on improved lithographic techniques. In Fig. 3, trends in feature size are plotted with predictive extrapolations in an effort to define the requirements for the 1980s. Fundamental differences will occur in patterning techniques in order to achieve the required feature sizes for VLSI. The shrinking feature size has also necessitated a reduction in film thicknesses, and vertical scaling has become a necessary consideration, not only to achieve the desired physical dimensions, but also to achieve the proper electrical parameters.

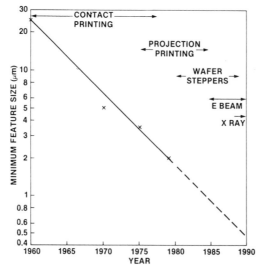

Fig. 3. Trend in minimum feature size and imaging technology.

The third aspect in Moore's thesis, contributions to increased circuit density through circuit design cleverness, poses additional challenges for the manufacture of VLSI. Quite often this cleverness results in increased complexity through the addition of circuit functions. For example, in the evolution of MOS technology to HMOS, the addition of devices of differing threshold voltages resulted in additional photomasking and ion-implant operations. Thus cleverness, in a sense, can be related to process length. Trends in process complexity, based on length, are illustrated in Table I.

TABLE I

Trends in Process Complexity

	Process					
Parameter	Al gate PMOS (1969)	Si gate PMOS (1970)	Si gate NMOS (1972)	Depletion NMOS (1976)	NMOS (1977)	NMOS (1979)
Gate length (μm)	20	10	6	6	3.5	2
Junction depth (μm)	2.5	2.5	2.0	2.0	0.8	<0.8
Gate oxide thickness (Å)	1500	1200	1200	1200	700	400
Mask levels	5	5	6	6	7–10	7–10
Diffusions	3	4	5	5	7–9	7–9
Relative complexity	1	1.2	1.3	1.7	2	2.1–3

Based on the evolutionary trends described, one can postulate the changes anticipated in manufacturing process technology for VLSI integrated circuits:

(a)   improved lithographic techniques,
(b)   increased use of dry processing,
(c)   lower-temperature operations,
(d)   more complex processes,
(e)   larger-diameter wafers,
(f)   more sophisticated facilities, and
(g)   higher-quality materials.

These items will be discussed in more detail throughout this chapter.

III. HORIZONTAL SCALING

In this section the horizontal scaling techniques required to implement VLSI circuit functions through cost-effective manufacturing processes will be reviewed. While more complex circuit designs and larger die will both evolve, horizontal scaling will continue to be the dominate method in achieving more functions per chip. The ability to achieve minimum feature size with smaller pitches is the objective of the patterning process. Feature delineation techniques and etching techniques to achieve the desired feature size will be described.

Historical trends in the minimum resolvable linewidth are shown in Fig. 3 along with the imaging technique used to achieve that level of resolution. Projected dimensions for VLSI circuits, along with alternatives for achieving them, are also given in the figure.

A. Present Capability

The processes used for pattern definition have developed over the years from contact printing of negative resists with wet chemical etching to a transition phase with projection printing, negative and positive resists, and wet chemical and dry plasma etching. Current tools and processes allow the printing of devices with 3-μm feature size on 100-mm wafers with ± 1.0-μm registration and 60-wafer/h throughput [7,8]. In the 1980s, extended use of projection printers, introduction of wafer-stepper and e-beam machines, and broader use of plasma processing seem apparent. It is also anticipated that the inner-mixing of various patterning techniques will exist. This change in manufacturing procedure will occur be-

TABLE II

Pattern Definition Trends

	Period			
Mode	1960s 24.0–6.0 μm Feature size	1970s 6.0–3.0 μm Feature size	Early 1980s 3.0–1.5 μm Feature size	Late 1980s 1.5–0.5 μm Feature size
Printing	Contact	Contact Hybrid contact/ projection Projection	Contact Projection Hybrid projection/ stepper	Projection Hybrid projection/ stepper e-Beam/x-ray
Resist	Negative	Negative/positive	Negative/positive	Positive
Etching	Wet chemical	Wet chemical Dry plasma	Wet chemical Dry plasma	Wet chemical Dry plasma

cause of the cost and complexity of the new machines as well as the widely varying process demands of the different operations. Table II illustrates some of the possible modes of operation for patterning VLSI circuits. All mask layers are not equally demanding for minimum resolution and alignment tolerance, and it is anticipated that mixing and matching various requirements to the proven capabilities of the various exposure tools will be common. For example, the alignment of contacts would be critical, requiring wafer-stepper or e-beam machines, whereas ion-implant isolations could be done with a projection printer or contact printer.

What then are the imaging alternatives for VLSI? The various approaches to fine-line lithography are shown in Fig. 4. Not all the alternatives shown are essential to achieving the dimensions required for VLSI implementation, and only those that appear likely candidates for production processes will be described.

B. Contact or Proximity Printing

Contact printing is the most established technique for transferring an image to a wafer. In this process an emulsion or chrome-surface mask is aligned and clamped directly on the surface of the wafer. Here the resolution is limited by the wavelength of the light (4000 Å) used to expose the photoresist. Improvement in the resolution can be achieved by extending the wavelength to the deep UV range (2000 Å). Machines using visible

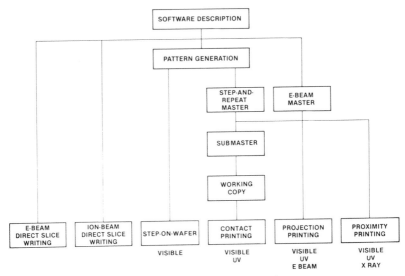

Fig. 4. Horizontal scaling alternatives.

wavelengths capable of achieving 1-μm feature size are available, while deep ultraviolet machines achieving 0.5-μm feature sizes have recently been introduced [9–11].

Registration accuracies of ± 0.125 μm are claimed for both contact printing and off-contact printing systems and, while the machines are capable of such accurate placement, the production difficulties of obtaining matched mask sets to ± 0.125 μm in sufficient volume to maintain a contact printing or an off-contact printing line is doubtful.

A major shortcoming of contact printing is the high level of defects introduced during the clamping process. This problem becomes more significant as the VLSI die becomes larger and the yield becomes defect limited, as discussed in Section V. Even though the feature size reproduction capabilities of these machines are adequate for the first generation of VLSI, it is unlikely that they will provide sufficient yield to be cost-effective in the long term.

C. Projection Printing

In projection printing the mask and wafer never contact one another, and the problem of defects encountered in contact printing is eliminated. Here the image is transferred to the wafer through a series of lenses or mirrors. Projection printers have been available for some time but were not initially required for the small die with large feature sizes. Projection printers were only successful in manufacturing after the development of a reflective, scanned, 1:1 system by Perkin-Elmer Corp. With this system, feature sizes less than 3 μm have been achieved in a production environment [7].

Registration accuracy of scanned, 1:1 systems has been quoted as better than 1 μm [7]. However, the total manufacturing accuracy achievable is down-graded in this type of machine by the absolute magnification errors encountered when one exposes various layers on different machines. Moreover, the alignment system in the 1:1 projection printer is a global alignment; two points on distant sides of the wafer are aligned and it is assumed that the rest of the wafer will register. These systems obviously depend on (1) the overlay accuracies inherent in the masks; (2) the die fit and die rotation errors being within the specified tolerances; and (3) the absence of process-induced wafer distortion. Wafer distortion alone can exceed 1 μm due to process-induced stresses and crystal damage [12]. Low-temperature processing and machine linear corrections may reduce wafer distortion to acceptable levels for VLSI. The extension of 1:1 projection printing to 1-μm feature sizes with alignment accuracies of ± 0.1 μm will require fundamental changes in the machine, in the facility in which it operates, and in the present methods of generating masks.

D. Wafer-Steppers

Many of the problems in both contact printing and projection printing, such as contact defects or registration inaccuracies, are overcome by wafer-steppers. These machines are an outgrowth of mask-steppers and project an image onto a portion of the wafer and then replicate that image across the wafer. These machines provide a minimum linewidth of 1.3 to 2.0 μm and a registration of 0.35 μm [13,14]. While these machines achieve excellent feature size and registration, they are relatively large, costly, and have low throughput. They offer the additional advantage over contact printing or 1:1 projection printing of defect size reduction. These 10× steppers project a 10× reduction of the die onto the wafer, and, therefore, dust particles on the reticle plate are reduced in diameter by a factor of 10. A 100-μm particle would print as a 10-μm particle; a 1-μm particle, which in 1:1 projection printing would kill a die, would not be resolvable in a 10× stepper. This, however, has no impact with regard to particles on the wafers that are responsible for many of the masking defects. It is generally believed that the 10× reduction capability of the current generation of wafer-steppers will allow consistently higher die yields to be achieved in a production environment. It should also be pointed out, however, that a single defect, when printed, could destroy all die on a wafer or in the run [13].

E. Electron Beams

In all of the prior techniques an image was transferred to the wafer through the use of a mask, which resulted in resolution limited by the wavelength of the exposing radiation. Direct electron-beam exposure of wafers is a technique that eliminates many of the problems associated with optical lithography. As seen in Fig. 4, e-beam imaging is conceptually the simplest of the imaging alternatives in that it translates the software description of the circuit directly to the wafer without any intervening process steps. Its greatest advantages are its minimum achievable linewidth of approximately 0.1 μm and its ability to achieve registration accuracy of ± 0.1 μm [15,16]. This high-resolution capability and registration accuracy make e-beam direct-wafer exposure an ideal tool for increasing circuit integration and packing density. The inclusion of fiducial markers on each die renders the technique insensitive to in-plane wafer distortion, a serious shortcoming encountered in 1:1 projection printing of die with small feature size. As the die becomes larger, mask defects become more significant in limiting yield (see Section V), and the use of e-beam direct-wafer writing minimizes the impact of these defects.

The most serious limitation to the use of e beams for direct exposure of wafers is the exposure cost, which results from the high machine cost and low throughput. IBM has reported a production machine with a throughput of 22 2$\frac{1}{4}$-in. wafers/hr [17]. The cost using e-beam machines has been examined by Henderson and by Piwczyk [18,19]. Piwczyk has found that it is more economical to use optical techniques for more than 2–3 wafers, but that for development activities, direct writing is cost-effective. Applying Henderson's model for circuits with an area exceeding 1 cm^2, e beam becomes cost-effective even for defect densities as high as 0.05 defect/cm^2. Based on these analyses, it is anticipated that direct-wafer writing will be limited to circuit prototyping in the near future.

F. X Rays

The use of x rays as a solution to the resolution limitation imposed on optical systems by defraction has been proposed as an alternative imaging technique for VLSI (see the review by McCoy [20]). Since the wavelength of the x rays is on the order of 10 to 50 Å, resolution of significantly better than 0.5 μm is achieveable. Although there is considerable development activity on x-ray imaging systems, a number of problems with alignment and masking still exist. Most of the x-ray systems are experimental in nature and it is unlikely that x-ray printing will be a production tool for VLSI in the 1980s.

G. Photoresists

Photoresists are light-sensitive organic polymers that are converted, after appropriate exposure and development, into etch-resistant patterns. They do not, in general, limit the minimum resolvable linewidth but are significant in the overall patterning process. Negative resists, whose image is the reverse of the mask image, do have a practical minimum space achievable. It is limited to about twice the film thickness because in the developing process the whole resist mass swells by absorbing developer solvent and adjacent geometries swell to the point of bridging. The main advantages of negative resists are lower cost, better adhesion, and faster photospeed; the primary disadvantage, relative to positive resists, is poor minimum resolution, which is about twice the film thickness.

Negative resists have been very successful in contact printing lines where the minimum resolution requirement has been about 5 μm. As the minimum resolution is being pushed toward 1 μm and projection printers are being used more frequently, positive resists must be used.

Positive resists require more exposure, provide generally poorer adhesion, and are much more expensive than negative resists. However, since there is no absorption of the developer by the unexposed resist, there is no resist swelling and, therefore, the practical resolution minimum can be well below the film thickness.

The resolution available in deep UV for positive resists will make them the primary resist systems for VLSI lithography. Moreover, the behavior of these positive resist systems has been remarkably well characterized; first by Dill *et al.* [21] and later by others [22,23].

While characterization of positive resists for optical lithography continues, the greatest need is for the development of higher-speed resists for both e-beam and x-ray lithography. The limitation of e-beam lithography has primarily been one of economics, contributed to by the slow speed and inadequate etch resolution of the resist. In addition, most e-beam resists are proprietary and unavailable for commercial application, although this is changing rapidly. Summary information is available in the articles by Martel and Thompson [24] and by Elliot [25].

In comparison, x-ray resists are even less characterized than e-beam resists and are too slow for production [25]. Hause *et al.* developed a 1-μm x-ray process using the negative resist COP, poly(glycidal methacrylate-*co*-ethyl acrylate), which required exposure times of 10 min/layer [26]. It is anticipated, however, that as the x-ray process tools develop, resist improvements will keep pace since the benefits of x-ray lithography are expected to impact VLSI for geometries less than 0.5 μm.

H. Imaging Appraisal

The parameters described for each of the imaging tools previously discussed are summarized in Table III. The three overriding criteria for lithography to realize manufacturing capability for VLSI functions are feature size, alignment tolerance, and defect levels. Which of the imaging alternatives is utilized in achieving the requirements for VLSI depends on how quickly the respective tools are developed and how the lithographic processes are integrated into the entire process flow. It is evident from past trends that the combining of various tools to achieve specific requirements will increase. This trend will be enhanced due to increases in cost and decreases in throughput of the new tools. It is also apparent that e-beam imaging has a narrow window for acceptance, and its use as a production tool depends heavily on the speed with which x-ray lithography is developed.

I. Pattern Definition

Lithography alone does not determine the feature size in the final circuits. Rather, the size is determined by a combination of the patterning and subsequent etching. Wet chemical etching has been the mainstay of

TABLE III

Imaging Capability of Various Tools

Tool	Wavelength	Linewidth (μm)	Registration (μm)	Throughput (100 mm W/hr)
Hard Contact	UV	1	±0.125	60
	Deep UV	0.5	±0.125	
Proximity	UV	3		60
	Deep UV	2		
1:1 Projection	UV	3	±1.0	60
	Deep UV	~1.0	±1.0	
Steppers	UV	1.25	±0.35	25
X Ray		<0.2	1	
E Beam		>0.2	0.1	7

pattern definition but is gradually being replaced with plasma etching, first, of silicon nitride layers, subsequently, of polysilicon layers and oxide layers, and, finally, of interconnects. This shift from wet chemical etching to dry plasma etching is taking place because of the inherent capability, quality, and process control advantages of dry processing as required by VLSI manufacturing. First, dry etching processes are anisotropic, reducing undercutting and permitting patterning of smaller feature sizes. Second, minimal use of chemicals is required, reducing the likelihood of accident or misprocessing.

1. Plasma Etching

The physics and chemistry of plasma etching are described in detail in Volume 2, Chapter 1. In plasma processing an inert molecular gas is excited in an rf or dc electric field in order to dissociate a reactive species, which then reacts with the material to be etched. A volatile product is formed, which is removed from the reaction chamber by the vacuum pump. Plasma processing is preferable to wet chemical etching primarily because it can etch anisotropically and because it is, inherently, a cleaner process, although it cannot "wash off" particles as wet chemical etching can. There are, however, problems with the universal use of plasma etching to pattern thin films. The principal difficulties with plasma etching are (1) reduced etch rate selectivity; (2) reduced etch uniformity; (3) generally higher cost, especially where parallel-plate etchers are concerned; and (4) lack of availability of production-worthy commercial equipment. Whereas difficulties (1)–(3) will probably always exist with plasma etching, the availability of production-worthy commercial equipment has been drastically changing, and it is expected that by 1985 several processes and commercial etchers for patterning polysilicon, silicon oxide, and aluminum will be available [27–29].

2. Reactive Ion Etching

Reactive ion etching provides improved anisotrophy compared to plasma etching and is effected in specially configured plasma reactors [30–32]. The two primary differences between plasma etching and reactive ion etching are the electrode configuration and the operating pressure. The electrode on which the wafers are supported in reactive ion etching must be considerably smaller than the ground electrode and must be capacitively coupled to the rf signal, providing electrical field lines normal to the wafer surface. The reactive ion species etch anisotropically since they follow the normal field lines [33]. Reactive ion etching is normally carried out at pressures lower than plasma etching. Polysilicon,

Si_3N_4, and SiO_2 have all been etched by reactive ion etching [34–36]. While this method provides improved anisotrophy over plasma etching, it is an exceeding slow process, and its extensive use in production in the near future is unlikely.

3. Ion Milling and Sputter Etching

Ion milling and sputter etching are distinct from plasma etching in that material is removed by purely mechanical means [37,38]. In sputtering, inert gas atoms are excited in an rf field and the momentum of the atoms knocks material off the surface. The etch rate in sputtering is extremely low; a problem that was solved with ion milling, in which a beam of ions is concentrated on the surface, increasing the etch rate. The selectivity of both sputtering and ion milling is poor, and the mask must be as thick as the pattern to be etched; however, the etched edge contour is nearly vertical with no undercutting. Ion milling can be used easily to define geometries down to 1.0 μm. In addition to the slow etch rate and poor selectivity ion milling has other processing difficulties, including wafer heating, redeposition of milled material, and secondary sputtering [39,40].

IV. VERTICAL SCALING

Vertical dimensions are critical for oxides, diffusions, and interconnects in achieving both the desired smaller horizontal dimensions and the proper electrical properties. It is easier to pattern small feature size in thin layers, especially with isotropic wet etches, but also with anisotropic plasma etches. However, the electrical impact of the thinner layers and dopings must also be considered in scaling. The thinner gate oxides and smaller contacts pose a potentially serious reliability problem. In this section, not only thin-film operations associated with vertical scaling but also other processes that impact VLSI manufacturing, such as high-pressure oxidation and ion implantation, are discussed.

A. Oxides

Gate oxide thicknesses had been relatively constant near 100 nm until the advent of horizontal scaling and the introduction of products on such technologies as HMOS [41]. Products are now available with gate oxides as thin as 40 nm, and it is reasonable to expect devices with thinner gate oxides in the future [41,42]. With such thin oxides, gate breakdown becomes a potential yield limiter. Oxides grown in the presence of chlorine have superior properties to those grown in a dry environment and are likely to be required for thin gate devices. The breakdown voltage of chlorinated oxides has been studied and shown to have a tighter distribution and higher average breakdown [43]. In addition to higher average breakdown voltages, material with chlorinated oxides has exhibited increased minority carrier lifetime [44]. While the trend in gate oxides is toward thinner layers, field oxides are remaining relatively constant.

The long, high-temperature, field oxidation in silicon devices presents a problem of compatibility with VLSI processing. High-pressure oxidation offers a technique for growing silicon oxides at reduced temperature and times. Initial characterization of high-pressure oxidation began in the 1960s [45], but it was not until recently that device applications have been noted [46].

The initial process benefits anticipated with the formation of field oxides via high-pressure oxidation of silicon included a significant reduction in oxidation time or temperature for a given thickness oxide when compared to equivalent conditions at atmospheric pressure. Steam, dry, and pyrogenic oxides have been grown at elevated pressures. It is possible to grow dry field oxides at elevated pressure more quickly than steam oxides at atmospheric pressure, thus improving the overall quality of the field oxide. It has been demonstrated as well that the fixed charge and interface density of high-pressure oxides are comparable to those of conventional oxides [47,48].

The initial benefit anticipated was the low-temperature capability associated with pressure oxidation, which would minimize warpage of large-diameter wafers and reduce unwanted diffusion. However, Tsubouchi et al. observed a reduction in density and size of oxidation-induced stacking faults in 6.4-atm pyrogenic steam field oxides when compared with equivalent atmospheric-grown field oxides and an improvement in dynamic RAM refresh time [46]. With the demonstrated benefits, commercial equipment has become available from Applied Materials Corporation, Thermco Products Corporation, and Tel-Thermco. These systems are typically capable of up to 25 atm, at temperatures between 600 and 1100°C with a capacity of 200 to 250 wafers.

B. Chemical Vapor Deposition

The direction of VLSI manufacturing toward thinner films to support horizontal feature size and the requirement for lower-temperature processing necessitated by larger-diameter wafers and less underdiffusion in-

TABLE IV

Deposition Conditions for CVD Films

Film	APCVD Temp (°C)	LPCVD		Plasma CVD	
		Pressure (torr)	Temp (°C)	Pressure (torr)	Temp (°C)
Si_3N_4	35	0.5	800	0.3–3.0	240–400
SiO_2	350–500	0.8	900	0.5–2.0	100–350
Polysilicon	650–1000	0.4	600	0.3–1.0	100–300

crease the need for low-temperature chemical vapor deposition (CVD) operations. The most commonly deposited films are silicon nitride, silicon dioxide, and polysilicon. These layers may be deposited by atmospheric pressure CVD (APCVD), low-pressure CVD (LPCVD), or plasma-assisted CVD [49–51]. The conditions under which deposition occurs are listed in Table IV. For VLSI circuits it is advantageous to shift from APCVD to LPCVD. Plasma-assisted CVD has advantages and disadvantages, and its use for a particular process must be evaluated on its own merits.

In general, LPCVD systems provide significant advantages for VLSI processing. They are typically operated at lower temperatures than thermal or APCVD systems, which reduces impurity redistribution and minimizes warpage. When compared with APCVD systems, LPCVD systems minimize defects due to particulates since the wafers are vertical rather than horizontal and since the film deposited on the chamber wall adheres better to a hot wall than a cold one. Vertical stacking of wafers in a boat increases throughput approximately fourfold over horizontal reactors. Temperatures are more uniform and easier to control in an LPCVD system resulting in improved uniformity ($\pm 3\%$ versus $\pm 5\%$) and step coverage. Low gas flows in an LPCVD system result in lower input costs. Plasma CVD systems achieve the lower-temperature operation but suffer seriously from reduced throughput because of the load configuration.

C. Source–Drain Technology

The necessity of using short channel lengths to achieve the small cell size required by VLSI and for low Miller capacitance to achieve the speed requirements point to a need to examine source–drain technology. Arsenic source–drain regions have supplanted phosohorus in n-channel silicon gate MOS devices, owing to arsenic's high solid solubility and low diffusion constant. Several techniques for forming the arsenic source–drain regions are available, including arsenic vapors in an enclosed capsule, an arsenosilica film, and ion implantation. Using the enclosed capsule technique, a surface concentration of approximately $3 \times 10^{20}/cm^3$ can be achieved at 1000°C, while for arsenosilica film sources, surface concentrations of approximately $1.6 \times 10^{20}/cm^3$ can be achieved [52,53].

The use of ion-implanted arsenic source–drain junctions for n-channel MOS integrated circuits has been described by a number of authors [54–56]. Implanting, both directly into the silicon and through oxides into the source–drain region, has been described using doses of 2 to 4 × $10^{15}/cm^2$ and energies up to 200 keV. To use this technology for doping source–drain regions of integrated circuits, major changes in the implantation equipment must be made in order to be cost-effective when compared to doping from a closed capsule or an arsenosilica film. With most currently available machines, the throughput for such high dose implants is inadequate for a manufacturing process and new machines are being developed. Table V lists ion-implant parameters for low-, medium-, and high-current machines.

Even though the use of ion implantation to dope source–drain regions has been discussed for device applications, some technology and equipment issues are unresolved. Wafer heating [57,58], charging effects [59], and neutral ion contamination all cause process-related problems. Low- and medium-current machines have used electrostatic beam scanning, while high-current machines have required the development of mechanical scanning systems to overcome beam space charge and wafer heating [58–60]. The mechanical scanning systems are complex and present serious maintenance problems.

Process problems associated with changing etch rates of nitride films exposed to high-current ion beams [61], increased reoxidation growth

TABLE V

Implant Parameters for Low-, Medium-, and High-Current Tools

Current	Source	Scan	Current limit	Cost	Energy (keV)
Low	Cold cathode	X–Y electro-static	To 100 μA	$250K	25–200
Medium	Hot fila-ment	X–Y electro-static, hybrid	To 600 μA	$275K	25–200
High	Hot fila-ment	Mechanical	4–10 mA	$500–600K	To 100 To 200

rates in implanted areas [62], anomalous arsenic profiles in deep substrate regions [63], and the inability to use photoresists as an implant mask all must be overcome to use high-current implanters. Charging effects associated with high-current implants have caused premature breakdown of insulating layers [64].

To overcome the throughput and process problems, high-current production machines are in development at Nova, Extrion, Lincott, Balzers, and Accelerators, Inc. Specifications for all these systems include mechanical scanning and provide currents up to 10 mA. The maximum energy and throughput capability vary among the manufacturers and must be considered for the specific machine application. For example, a process alternative for ion-implanted source–drain regions for VLSI circuits is whether implanting will be done through an oxide or directly into the silicon. Both machine and process development are sensitive to this alternative.

On (111)-oriented silicon and to a lesser extent on (100)-oriented silicon, bulk mobility degraded in through-the-oxide-implanted regions owing to oxygen recoil [65]. Energy requirements are also different for through-the-oxide-implanted source–drain regions, and some machines will likely be developed providing implant energies less than 100 keV and others to 200 keV.

D. Interconnects

The use of refractory metals, monosilicides, and disilicides for gates and/or interconnects has long been investigated, initially as a replacement for the aluminum gate [66,67]. The extensive use of silicon gate and polysilicon interconnects, however, slowed the development of refractory disilicides for interconnects. For VLSI circuits, not only processing but also circuit considerations are increasing the emphasis on the development of these materials as an alternative in order to overcome the limited conductivity of 10 to 20 $\Omega/\square$ associated with typical 5000-Å-thick doped polysilicon lines. As circuits are scaled by a factor k, interconnect resistances increase by a factor of $1/k$ and contact resistance by $1/k^2$. Longer and thinner interconnects may require lower resistivity materials to provide reasonable access time, as illustrated in Fig. 5. The cost of using such interconnect systems is increased process complexity, and circuit design cleverness may provide alternative techniques in achieving the improved access times.

Refractory metals and their disilicides are ideal candidates for VLSI interconnects because they offer the low resistivities and high melting point

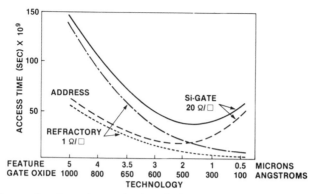

Fig. 5. Access time for various technology limits using polysilicon and refractory metal interconnects (after Shah [67a]).

required for silicon device processing. Which of the disilicides is most promising is still not determined but a number of them can be eliminated owing to their relatively low melting points ($PtSi_2$, $PlSi_2$), poor etching characteristics, and, in some cases, high resistivity (CrSi, ZrSi). Those refractory metals and their disilicides that have been suggested as most promising interconnects are molybdenum, titanium, tantalum, and tungsten. In Table VI, a list of properties of these refractory metals and their disilicides is given. In addition to compatible physical properties, several of the disilicides are close to polysilicon chemically and, therefore, the use of refractory metals and disilicides in manufacturing VLSI is not significantly different from the use of aluminum or silicon. Disilicides of both

TABLE VI

Properties of Some Refractory Metals and Their Disilicides

Material	Melting point (°C)	Electrical resistivity ($\mu\Omega$-cm)	Primary oxide
Si	1420		SiO_2
Al	660	2.8	Al_2O_3
Mo	2620	5.3	
Ta	2996	13.1, 15.5	
Ti	1690	43, 47	
W	3383	5.3	
$MoSi_2$	1870	21.5	MoO_3
$TaSi_2$	2400	8.5	Ta_2O_5
$TiSi_2$	1540	16.1	TiO_2
WSi_2	2050	12.5	WO_3

TABLE VII

Processing Conditions for Devices Fabricated with Refractory Gates[a]

Device	Gate	Deposition	Anneal	Etch	Thickness (Å)	Resistivity	Reference
4K SRAM	Mo	e-Beam	Forming gas	N/A	N/A	N/A	[68]
64K DRAM	Mo/Si	N/A	Forming gas	N/A	2000	10^{-5} Ω-cm	[69]
16K DRAM	Mo	N/A	N/A	N/A	N/A	N/A	[69a]
MOSFET	MoSi$_2$	rf Sputtered	Hydrogen	$CF_4/4\%$ O_2	3000	2 $\Omega/\square$	[70]
Capicators	WSi$_2$	rf Sputtered	Argon	2% HF/98% HNO_3	2500	10^{-5} Ω-cm	[72]
Capicators	TaSi$_2$	Magnetron	N/A	N/A	2500	2.2 $\Omega/\square$	[72a]
Capicators	TiSi$_2$	Cosputtered	N/A	N/A	2500	0.9 $\Omega/\square$	[72a]
MOSFET	WSi/Si	Coevaporation	Hydrogen	RIE	2500/1500	2.6 $\Omega/\square$	[73]

[a] SRAM, static random access memory; DRAM, dynamic random access memory; N/A, not available.

tungsten and molybdenum are soluble in an $HF + HNO_3$ wet etching solution.

The refractory metals can be used for interconnects alone, as a disilicide, or as a composite structure. Recently, devices have been fabricated using molybdenum gates and interconnects [68,69], molybdenum silicide gates and interconnects [70,71], tungsten silicide gates [72], and a multilayer structure of tungsten and polysilicon [73]. A summary of the processing data for the various devices is given in Table VII. Refractory disilicide films are formed by e-beam evaporation, rf or magnetron sputtering from hot pressed targets, and chemical vapor deposition [68,70]. Each method requires care during deposition to avoid contamination [74].

The films may be codeposited or directly deposited by putting the refractory metal onto silicon and heat treating it. Cosputtering has the advantage of using two pure, elemental targets for the source. In these films, which are cleaner than sputtered films, it is more difficult to control the stoichiometry. The films can be patterned by wet chemical etching [71], plasma etching [70,71], or ion milling. For all refractory disilicides, annealing is required to achieve the optimum resistivity, and improvement of an order of magnitude over doped polysilicon can be achieved.

The use of refractory disilicides as gate materials requires good contact resistance and reliability. The $Al/MoSi_2$ system has been well characterized and adequate contacts are formed between $Al/MoSi_2$ and $MoSi_2$/doped polysilicon [65–75]. Reliability and corrosion resistance of a number of interconnect systems have been reviewed [76]. $MoSi_2$ will decompose in the presence of moisture during or after processing. Wafers must be dry prior to furnace treatment of an exposed $MoSi_2$ film and the films must be coated with high-quality dielectric films to prevent subsequent corrosion.

V. CHARACTERIZATION AND PROCESS CONTROL

Characterization and process control techniques must be improved significantly to achieve acceptable yields for VLSI circuits as feature size decreases and die size increases. The simple use of in-process visual monitors, thickness measurements, test patterns, and failure analysis after electrical tests must give way to more advanced and faster schemes for determining the performance of a particular process operation. A composite manufacturable process for VLSI implies that a sufficient fraction of the die on a wafer meet the electrical specifications. This fraction, expressed as a percent, is called the die yield. A second yield, consisting of that fraction of the wafers that started in the process and finished, is termed *line* or *process yield*. The product of the two is *overall yield*. Characterization and process control are used to maximize the overall yield. The die yield is generally considerably lower than the line yield, and most characterization and process control are applied to improving it. Die yield losses are due either to parametric failures or functional failures. On well-characterized and mature products, functional failures tend to dominate and can usually be attributed to processing defects. These defects are related either to the facility or to specific process steps. The parametric failures, as well, can generally be related to a particular process operation. It is the function of characterization to identify the process sensitivities and of process control to monitor them at the respective operation.

A. Defect-Limited Yield

Effective models for defect-limited yield exist and are described in the literature by Price [4], Stapper [6], and Lawson [77], and others [78–80]. In all the models, the decreasing die yield with increasing die size is explained based on statistical defect-density distributions. The authors differ in the statistics applied and the results of their calculations are plotted in Fig. 6. Price [4] uses Bose–Einstein statistics to derive an ex-

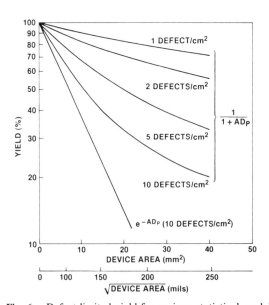

Fig. 6.   Defect-limited yield for various statistical models.

pression for the yield Y in terms of the die area A and the process defect density D_p as

$$Y = 1/(1 + AD_p).$$

Lawson uses Boltzmann statistics to derive

$$Y = \exp(-AD_p),$$

while Stapper uses a mean defect density D_m and standard deviation σ to derive

$$Y = (1 + AD_m\sigma^2)^{-1/\sigma^2}.$$

In reality these descriptions apply only to a single process step, but in practice they have successfully been applied to an entire process flow. While there is some discussion over which of these is exactly correct, the features are the same.

In moving toward VLSI, three trends are occurring in design and process development that, when examined in the context of the preceding equations, suggest problems for the manufacturer of VLSI circuit: (1) the area A is increasing, as seen in Fig. 2; (2) the number of levels N is increasing as seen in Table I; and (3) the feature size is decreasing and hence the number of effective defects D is increasing.

All of these trends tend to lower the yield for a given process capability, and, for the larger die, small variations in yield have a significant impact on cost. For this reason, increasing use is being made of process analysis structures to identify yield-limiting process steps [81,82]. These test structures are not only being used to develop high-yield processes but are also used as in-process monitors [83]. Such monitors are capable of providing a quantitative distribution of defect density versus defect size for a given process operation.

Evolutionary changes in manufacturing facilities and equipment are occurring to reduce defect density in the facility. Slices are kept in a clean environment throughout the process, particle-free materials are being introduced everywhere, and filtered chemicals and central piping systems are being utilized.

Revolutionary changes in design, such as the use of fault tolerance, are being applied to overcome the negative impact of VLSI trends in manufacturability. Failure analysis of larger memory circuits suggests that a majority of the functional failures may be attributed to small physical defects resulting in single-bit, row, or column failures. Designers, recognizing this, have considered including spare elements in the circuit layout, which could be substituted for defective elements [84–89]. While this technique increases the die size, it has the potential of increasing the

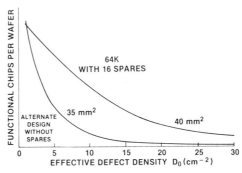

Fig. 7. Impact on fault-tolerancy on 64K DRAM yield (after Cenker *et al.* [84]).

yield. Cenker et. al. have estimated the impact of the yield of incorporating fault tolerance in a circuit design and their results are in Fig. 7. This technique has the potential of changing the nature of defect-limited yield statistics as they relate to integrated circuit manufacturing.

B. Process Control

The techniques for process control during the manufacture of integrated circuits have not developed at the same rate as the circuits and equipment innovations have not kept pace with the requirements of VLSI. The most often characterized parameters during IC manufacture are film thicknesses, resistivities, critical dimensions, and defect levels. Instrumentation exists for measuring both thickness and resistivity adequately, but significant individual judgment is required to characterize critical dimensions and defect levels. While there are several instruments available to measure critical dimensions, their reproducibility and calibration are inadequate for the circuits of the 1980s. At best, the instruments are capable of ± 0.1 μm and, when operator error is included, this increases to ± 0.2 μm over long periods [90]. When this is added to the process variability, it is evident that characterization and control of less than 1-μm gates will be impossible with existing tools. For VLSI manufacturing to succeed in production, improved equipment and techniques will be required for measuring critical dimensions.

Linewidth measuring equipment currently available uses either optical imaging methods with a measuring eyepiece or optical nonimaging techniques. The accuracy of these techniques depends on the optics, illumination, and edge detection criterion [91]. Imaging tools are made by Vickers Instruments Inc, ITP, Inc., Royokosha, and Nanometric Inc. These systems are capable of measuring down to 1 to 2 μm with 0.10-μm reproducibility.

Monitoring of defect levels throughout the process, which is critical to achieve acceptable yields, is the least automated and most subjective process control technique used in manufacturing. Operators inspect the individual dies under microscopes for visible defects. While no progress has been made in the automation of inspection techniques for wafers, there now exist machines for the automatic inspection of photomasks, which will be discussed in Section VI. It is necessary that this automated type of inspection equipment be extended to wafers.

In-process visual inspection of dies for defects is subjective, and operator turnover introduces a significant variable in the data base. Use is being made of special masks that, when processed through individual photolithographic operations and subsequently tested, provide quantitative measures of defect density for yield improvement. Similar techniques have been used to define optimum design rules for given process operation [81,82] and to monitor the performance of the wafer fabrication facility [83–92]. A test structure of automatic characterization of registration between two levels to better than 0.05 μm has been described [93].

VI. MATERIAL

A. Silicon

Considerable research has been performed in an effort to understand the relationship between material properties and device characteristics. The material requirements for VLSI circuits are complicated owing to (in spite of extensive materials characterization) a complex interaction between defects in the starting material and the subsequent processing. The material characteristics required are defined by the circuit's electrical requirements, the process sequence, and the productivity constraints of manufacturing.

As designers continue to shrink devices resulting in shorter channel lengths and the threshold voltage drops, the bulk resistivity must increase in order to prevent electrical punchthrough from source to drain. In addition, the smaller feature sizes that must be printed require flatter surfaces at each patterning operation. The starting material may be sufficiently flat, but successive thermal operations induce strains in the wafers, causing warpage, which results in distortion during patterning. The process developer, in conjunction with the designer, add complexity to achieve additional circuit functions as well as desired circuit properties. This complexity often results in additional thermal operations for annealing implants or gettering impurities tending to warp the silicon wafers.

Finally, the manufacturer, in an effort to improve productivity, increases the diameter of the wafers. Thus the critical silicon material thrusts for VLSI circuit fabrication are to: (1) continually increase diameter; (2) increase resistivity; (3) effectively use impurities; and (4) provide material that is physically stable during processing.

1. Mechanical Parameters

The silicon material requirements can be broadly divided into two categories: mechanical parameters (e.g., flatness, diameter) and physiochemical parameters (e.g., defects, resistivity). In Fig. 8, the trends in the diameter and area of silicon wafers are plotted versus time. The most cost-effective transition occurred between 2- and 3-in. wafers where an effective capacity and productivity increase of 125% was realized. In the industry's most recent conversion from 3-in. to 100-mm wafers, a 78% increase in area was realized. A conversion from 100 to 125 mm results in only 51% increase in capacity. Whether such a conversion would be cost-effective must be determined through detailed financial analysis. The capability of the silicon industry to produce large-diameter wafers now leads the semiconductor industry in its ability to process them. The use of still larger diameter wafers, in an effort to maintain productivity as die sizes increase, presents three major problems for the semiconductor manufacturer. First, an enormous amount of capital is required to retool equipment; second, changes in patterning techniques are required to print the smaller feature sizes over the larger area; and third, the need to main-

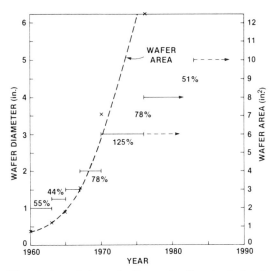

Fig. 8. Historical trends in production wafer diameter (—) and area (×).

tain flatness requires the development of low-temperature processing, a task that becomes more acute as processes become longer. Therefore, in the next transition to a larger-diameter wafer, a close relationship between the material manufacturer and process engineer will be essential, for even if the transition is proven cost-effective from an asset evaluation, the technical problems that must be overcome will be tremendous.

2. Warpage

Warpage is critical primarily because it affects the ability to define the required geometries throughout the VLSI manufacturing process. Since thermal stress, which leads to warpage, is a cumulative effect, the warpage gets progressively worse throughout the process, and it can be particularly severe at contact and metal patterning. When the stress in the wafer exceeds the yield point of silicon, dislocations form and plastic deformation occurs to partially relieve the stress. When the wafer has been cooled to room temperature, a reversed stress distribution, which cannot be relieved by plastic deformation, will be present. This stress causes the wafer to buckle [94], destroying its planar shape, and leads to photolithographic problems. The parameters that influence warpage the most are:

   (a)   the growth method of the starting material;
   (b)   the diameter and thickness of the wafers (especially the wafer diameter over thickness ratio);
   (c)   the nature of the cumulative thermal processing (temperature, temperature gradients, heating and cooling rates, configuration);
   (d)   the amount and form of the precipitates (oxygen) and the direction and magnitude of the initial bow; and
   (e)   the nature of the films grown or deposited on the wafer's surface [94–97].

The stresses that lead to warpage originate from temperature transients, which occur during the insertion of wafers into and the extraction of wafers from furnace tubes used in high-temperature processing. A difference exists between rapid heating and rapid cooling in the maximum temperature gradient achieved and in the temperature at which the gradient occurs. Since the larger gradient occurs at a higher temperature during cooling and since the elastic yield limit drops substantially as temperature is increased, it is the large thermal gradient reached at high temperature during cooling that causes slippage. Stresses increase as the wafer diameter-to-thickness ratio increases, and it is necessary to lower the temperature to remain below the elastic limit for large-diameter wafers.

The rapid cooling in the initial period of the pulling operation is primarily controlled by radiation leading to a radial temperature gradient in the wafers. The wafer periphery cools faster than the center, causing a temperature gradient of about 150°C, and creates thermal stresses in each wafer. These stresses can be large enough to exceed the yield stress of silicon and, consequently, generate dislocations. The magnitude of the stresses can be estimated by $\sigma = \alpha E \, \Delta T$, where σ is the thermal stress, $E \cong 1.5 \times 10^{12}$ dyn/cm^2 is the average Young's modulus, α the coefficient of linear expansion $\cong 4 \times 10^{-6}$ K. Thus, a thermal gradient of 150°C is already large enough (0.9×10^9 dyn/cm^2) to exceed the yield stress (0.45×10^9 dyn/cm^2) of silicon and to introduce dislocations.

The thermally induced dislocation will partially relive the thermal stresses. Since thermal stresses are only partially relieved, a reversed stress distribution arises in a wafer after cooling, which now, with the wafer at room temperature, cannot be relieved by plastic deformation. The wafer, being a thin plate, will relieve such strains by buckling or warping. Initial bows, impurities, precipitates, and heat cycles will play a major role in the warping process.

Substantial efforts to prevent warpage are required to pattern VLSI circuits. Gegenwarth and Laming have shown that plastic deformation of silicon wafers can contribute to lateral wafer deformation of up to 0.5 μm, which can have significant impact on layer-to-layer registration [98]. Variables such as temperature, temperature gradient, position of the wafers in the row, distance between the wafers, wafer diameter and thickness, and initial bow all have a great effect on warpage and must be optimized for each process to minimize the warpage. The use of slow push/pull rates and closed boats have been found to decrease warpage dramatically. Impurities also significantly affect the mechanical properties of the silicon, and differences in behavior between materials grown by the floating zone and by Czochrolski techniques are observed.

3. Gettering

The silicon wafers used as the starting material in VLSI circuit fabrication contain a variety of defects, many in the supersaturated state. Thermal processing of the silicon during wafer fabrication results in a redistribution of the grown-in and process-induced impurities and defects through diffusion. The high diffusivity and the changes in the solid solubility limits with temperature make the impurities, self-interstitials, and vacancies very susceptible to precipitation after thermal treatment [99]. Generally, the precipitation will occur preferentially at lattice-deformation sites.

The gettering technique involves the generation of lattice damage on the back surface (extrinsic) or in the bulk (intrinsic) followed by a thermal treatment. At high temperatures, the impurities diffuse to the strained region and precipitate at the lattice damage, away from the active area of the device. Gettering is required in the manufacture of VLSI circuits to eliminate the deleterious effects of contamination and process-induced defects. For example, transition metal precipitates in the junction region causes the soft reverse characteristic of the junction [100,101]. Other defects such as decorated dislocations and stacking faults also degrade the yield and speed of the devices. Thus, gettering involves the generating of sinks for the point defects at some location of the wafer away from the active region and then the annealing of the wafer at high enough temperature so that the point defects will quickly diffuse to the sinks. A variety of gettering techniques such as ion implantation [102–104], glass-layer doping [105,106], intrinsic (oxide) precipitation [107,108], and phosphorus gettering [109,110] have been used. Table VIII summarizes some of the gettering mechanisms that have been used [111].

TABLE VIII

Gettering Mechanisms for VLSI Processing[a]

Type	Mechanism	Comments
Physical (Cottrell)		
Backside ion implant	"Metallic" impurity incorporation in shallow dislocation	Implant damage may be annealed out
Backsurface abrasion	"Metallic" impurity incorporation in incoherent dislocation array	Irregular surface damage may propogate throughout wafer
Backsurface Si_3N_4	Macroscopic strain-enhanced impurity movement toward backsurface	[112]
Intrinsic	Local clustering of "metallic" impurities in SiO_x precipitate-induced dislocation and/or extended point-defect complexes generated through sequential thermal process	Extremely structure sensitive to thermal history of silicon material
Impurity–impurity		
HCl	Suppression of oxidation-induced stacking faults (surface)	[113]
Carbon–phosphorous–gold	Multielement interaction	[114]
Solubility-enhanced		
Phosphorus (PSG)	Formation of metal acceptor–phosphorus donor pair and/or misfit dislocation gettering	[110]
Isolation diffusion	Highly stressed silicon lattice	[115]

[a] Huff [111].

Since gettering means the intentional introduction of lattice deformation and the subsequent annealing at high temperatures, the manufacturer has to realize that these lattice deformations can also be detrimental. When a gettering technique is chosen, the following parameters have to be accounted for: (1) effectiveness of gettering; (2) type of lattice deformation; (3) density of lattice deformation; (4) relationship between the lattice deformation and the process steps; and (5) stress distribution induced by the growth or deposition of thin films and its effect on the lattice deformation.

B. Photomasks

The requirements for and nature of photomasks will change significantly in order to support VLSI circuit manufacturing. For projection printing, quality and tolerances will have to improve to cost-effectively manufacture the larger circuits with smaller feature size. Thus, e-beam-generated masks will become more prevalent as masters used for projection printing, since they provide better resolution and linewidth control, fewer defects, and faster turnaround [116]. The use of wafer-steppers for patterning will require reticles instead of master masks, shifting the emphasis of mask-making significantly. The need for photomasks will be altogether eliminated by the use of e-beam direct slice writing.

The photomask parameters that are significant to the manufacturer of integrated circuits are: (1) defect levels; (2) level-to-level registration; (3) minimum geometry size; (4) test site registration and sizing; and (5) expansion coefficient of the mask plate.

1. Defect Levels

Once the proper pattern sizes are defined, the defect density is the most critical mask parameter to the circuit manufacturer. The conversion from contact printing to projection printing, while reducing the minimum resolvable linewidth, improved mask quality significantly through the elimination of the working plates and conversion to the use of stepped master plates.

The discussion of defect-limited yield in the previous section applies to mask defects as well. Mask defects can be introduced during the manufacture of the mask or during use in subsequent processing. The mask-limited yield for a nine-level process as a function of die size for various defect levels is shown in Fig. 9. Inspection and cleaning techniques for masks are critical to achieving high yield on large dies. The dimension of a critical defect has decreased along with the decreasing minimum feature

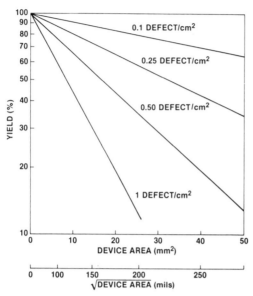

Fig. 9.   Mask-limited yield for a nine-level process with various defect densities per level.

size, and the need for lower defect density has increased with increasing die size. Until recently, manual optical inspection techniques have been adequate to monitor mask defect levels. Control was achieved through replacement of defective masks. However, the field size of an optical microscope at 200× is too small to be effective as a manual inspection technique for defect densities less than $5/cm^2$ and sizes near 1 μm. Operators are at best 50 to 80% effective under these conditions.

For these reasons, manual inspection techniques are giving way to automated inspection techniques [117–120]. The use of automated inspection has provided information for the improvement of a mask cleaning process resulting in improvement of defect density from approximately 1.5 to $0.5/cm^2$ over a one and a half year period [119]. The repair of masks and especially of reticles using spot exposure and laser burning of chrome will become more prevalent in achieving high-quality defect-free masks. Commercial inspection equipment has only recently become available. It is manufactured by KLA Instruments Corp., NJD Corp., Photodigitizing Systems Inc., and Stahl Research Laboratories Inc. These machines operate under different principles and can detect minimum defects ranging from 0.2 to 2.0 μm. While all the machines are expensive, the use of such equipment may be necessary to achieve reasonable yield on VLSI circuits.

2. *Registration*

Layer-to-layer registration becomes more important as feature sizes shrink. An incoming mask overlay is required to assure that the critical mask levels register to one another. A number of optical comparators for measuring registration are commercially available (Nikon Inc. and E. Lietz, Inc.) and have improved the circuit manufacturer's ability to assure registration of critical layers. The use of stepped master masks has greatly improved the layer-to-layer mask registration and the use of photorepeaters with laser-metered stages has resulted in masks with runout of ± 0.77 μm [117].

VII. THE MANUFACTURING FACILITY

The facility to implement VLSI will include not only the physical plant and support system but also the equipment and its configuration and the manufacturing and engineering systems. This comprehensive approach must be taken because of the strong relationship among the process, process equipment, and process environment. An example of this coupling is the photomasking operation. Defects in the environment (i.e., contaminants or vibration) or on the wafer or mask (i.e., particles) cause dead dies. Resist and exposure properties are sensitive to temperature and humidity. Increasing technological complexity creates the need for larger physical plants and more support equipment. The increase in dry processing, longer processes, rising cost of capital equipment, and larger-diameter wafers, all of which have been discussed earlier, will modify the requirements for building a semiconductor manufacturing facility for VLSI. Smaller feature size and larger wafers are having a dramatic effect on the nature of printing technology. In moving from contact printing, to scanned projection printing, to wafer-stepped printing, to e-beam direct-wafer writing, the cost of the capital equipment increases 10- to 50-fold as depicted in Fig. 10. Also shown is pattern etching equipment increasing in cost by the same amount. Changes in technology, replacing diffusion with ion implant, for example, create dramatic increases in equipment costs and facility-related expenses. Increased costs of facility-related items, such as clean rooms, power, analytical equipment, and test equipment, result in the exponential growth in the overall capital requirements for constructing a wafer-fabrication factory. The major share of semiconductor manufacturer's capital is invested in wafer fabrication and the requirements for VLSI are dramatically affecting the industry's revenue per capital dollar invested [121].

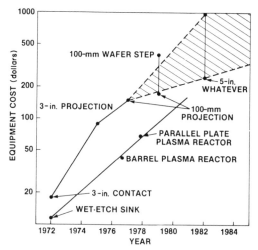

Fig. 10. Cost trends for pattern definition equipment (after Kauffman [121a]).

Semiconductor manufacturers will have to alter this trend toward increased costs while continuing to expand capability for VLSI. Mechanization and automation can dramatically impact the revenue per capital dollar invested by providing an environment for introducing VLSI products at higher yields and by increasing inventory turnover with shorter throughput time. The shorter throughput time will also aid in problem solving by reducing the impact of process-related yield excursions and in the rapid development of new products and technologies. Perhaps the greatest advantage to VLSI implementation realized through continued progress in automation will be reduced development throughput time.

The nature and extent of automation in a manufacturing facility depends on resources available. The automation can be categorized as mechanization, individual equipment automation, and integrated process automation. Complexity, cost, and process knowledge required to implement these levels of automation increase from mechanization to integrated process automation.

Mechanization provides two advantages in the manufacture of integrated circuits: first, it increases operator productivity; second, if done properly, it reduces handling and related visual defects. Examples of areas where mechanization have been applied are automatic flip transfer boat loading, automatic wafer scrubbers, and automatic loading and unloading of printers and ion implanters. These operations have no impact on the process or its control and affect only the mechanical nature of the operation.

Computers and microprocessors are beginning to appear in some pieces of stand-alone equipment such as wafer-handling equipment, diffusion furnaces, evaporation equipment, and printers. Diffusion furnaces, for example, now use digital computers to monitor and control furnace zone temperature, gas flow, boat motion and speed, system alarms, and operator requests.

Imaging equipment uses microprocessors for wafer loading and unloading, automatic aligning and focusing, reducing dependence on the operator, and increasing throughput.

Commercial equipment development that couples process operations through automation has been tried. However, this has been limited exclusively to the photolithographic operations. In this equipment, coat, bake, and develop operations are all coupled through an in-line system, with wafers transported automatically. None of the automation techniques have feedback capability from the prescribed process parameter to the machine for correction of operational parameters.

Conceptually, the direction of automation is toward complete coupling of process steps with feedback for parameter adjustment and with automatic management and engineering data collection. This automation, if done properly, can provide improved process control, shorter throughput time through the elimination of ques, and process flexibility. It has the disadvantage of high capital and engineering development costs.

The application of an integrated approach to an automated facility is considerably more costly than the construction of a conventional facility, and resources, in terms of people, limit the number of companies that can participate in this aspect of development. Semiconductor equipment manufacturers are particularly limited in their ability to support the automated facility development owing to their lack of process know-how. While photolithographic equipment manufacturers may understand the manufacture of printing equipment, their understanding of the requirements of the prior and subsequent processes and overall process flow is inadequate owing to the proprietary nature of the processes developed by the semiconductor manufacturer.

Therefore, to date, only the larger semiconductor manufacturers have had the expertise and resources to automate some of their manufacturing facilities completely, and while this automation will allow the fabrication of VLSI circuits with more control and lower defect densities, the driving force has been improved throughput times to reduce the design and development cycle.

IBM has recently described [122] a computer-controlled and highly automated single-wafer processing production line. The system consists of automated process sections into which wafers are loaded and processed while in a clean (Class 50) environment. The line can process single wafers and incorporates an automatic process control system, an electronic wafer routing system, and automated data collection and analysis, as well as many other features. A significant capability is the incorporation of feed-forward process control. An example of this process control capability is cited where SiO_2 film thickness is measured by a spectrophotometer and fed forward to an etching tool, permitting a process endpoint detection algorithm adjustment. While considerable software is utilized to provide process control and flexibility, one of the features emphasized was the ability to provide short throughout times.

Texas Instruments, Inc. has also described an automated, computer-controlled, VLSI, slice processing facility and a computer-controlled prototyping facility with fast turnover for VLSI design and development [123].

Complete factory automation will be required to provide capability, cost effectiveness, and support not only to manufacture but also to develop the VLSI circuits of the 1980s. The manufacturing tools are becoming more expensive and the process parameters more sensitive to external influences, resulting in a need for feedback and control systems, both for manufacturing and engineering data collection.

VIII. SUMMARY

Manufacturing process technology for VLSI was reviewed based on the thesis that density will increase through reduced feature size, increased die size, and increased complexity. It is clear that the manufacturing of large-scale integrated circuits will be achieved through continual evolution in manufacturing processes. How rapidly VLSI becomes a manufacturing reality will be limited only by how quickly the evolutionary trends can be synthesized into a process flow, since designers have sufficient knowledge to design circuits with 1.0 μm feature sizes. The speed with which these designs can be verified, however, is limited by the extent of design automation and the throughput time required to process verification.

The primary thrusts in manufacturing technology, which were discussed throughout the chapter, focused on continuing feature size and defect density reduction. To achieve these reductions, significant changes will occur in patterning tools and technology, while the remaining process operations will move toward lower-temperature processing, more extensive use of dry process, both for etching and deposition, and increased automation. As the equipment becomes more costly, evolving from contact to projection to stepped projection to x-ray or perhaps e-beam lithography, single-printing technology will no longer be utilized for all masking layers. The increased use of automation will appear first on individual tools and, later, for entire processes as efforts are made to increase productivity, cost effectiveness, and throughput time.

To satisfy electrical requirements and provide design flexibility, processes will become longer, necessitating improved process control techniques. There is no reason to expect that the trends projected in Fig. 1 will change as the industry moves to VLSI in the 1980s.

ACKNOWLEDGMENTS

The author wishes to acknowledge Jacob Aidelberg for his helpful suggestions on materials for VLSI, Moshe Balog for his contribution to the section on CVD, Walter Matthews for his comments on lithography, and John Orton for his help on ion implantation. John Caywood and Bob Jecman read the manuscript and checked for consistency. Karen Kieffer provided literature search service.

REFERENCES

1. G. E. Moore, Progress in digital integrated electronics, in IEEE International Electron Devices Meeting Tech, Dig., p. 11–13. Washington D.C., 1975.
2. R. W. Keyes, *IEEE Trans. Electron Devices* **ED-26**, 271 (1979).
3. B. Hoeneison and C. A. Mead, *Solid-State Electron,* **15**, 819 (1972).
4. J. E. Price, *Proc. IEEE* **58**, 1290 (1970).
5. C. H. Stapper, Jr., *IEEE J. Solid-State Circuits* **SC-10**, 537 (1975).
6. R. M. Warner, Jr., *IEEE Trans. Electron Devices* **ED-26**, 86 (1974).
7. M. C. King, *IEEE Trans. Electron Devices* **ED-26**, 711 (1979).
8. J. Dey, 1 to 3 μm lithography: How? *Proc. SPIE* **135**, 83 (1978).
9. K. G. Clark and K. Okutson, *Microelectronics* **6**, 51 (1975).
10. R. C. Heim, Practical aspects of contact/proximity, photomask/wafer exposure'', *Proc. SPIE* **100**, 104 (1977).
11. J. Lyman, *Electronics* **52**, 105 (1979).
12. L. D. Yau, *IEEE Trans. Electron Devices* **ED-26**, 1299 (1979).
13. J. Roussel, Step and repeat wafer imaging, *Proc. SPIE* **135**, 30 (1978).
14. W. C. Scheider, Testing the Mann type 4800 DSW™ wafer stepper™, *Proc. SPIE* **174**, 6 (1979).
15. M. Hatzakes, Lithography processes in VLSI circuit fabrication, *in Scanning Electron Microsc., 1979,* p. 275 (1979).
16. H. C. Pfeiffer, *IEEE Trans. Electron Devices* **ED-26**, 663 (1979).
17. R. D. Moore, Reliability availability, and serviceability of direct wafer exposure electron beam systems, *Proc. Int. Conf. on Microlithogr., Paris, France* p. 153 (1977).
18. R. C. Henderson, Direct wafer exposure using electron beam lithography vs. mask replication: A cost comparison model, *Proc. Int. Conf. Symp. Electron Ion Beam Sci. Technol., 7th* p. 205 (1976).

19. B. P. Piwczyk, An economic view of electron beam lithography, *Proc. SPIE* **100**, 120 (1977).
20. J. H. McCoy, X-ray lithography for integrated circuits, A review, *Proc. SPIE* **100**, 162 (1977).
21. F. H. Dill, W. P. Hornberger, P. S. Hauge, and J. M. Shaw, *IEEE Trans. Electron Devices* **ED-22**, 445 (1975).
22. F. H. Dill, A. R. Neureuther, J. A. Tuttle, and E. J. Walker, *IEEE Trans. Electron Devices* **ED-22**, 452 (1975).
23. J. Shaw and H. Hatzakis, *IEEE Trans. Electron Devices* **ED-25**, 425 (1978).
24. G. W. Martel and W. B. Thompson, A comparison of commercially available electron beam resists, *Kodak Microelectron. Sem. Proc., San Diego, California* (October 1978).
25. D. J. Elliott, *Semicond. Int.* 61 (May 1980).
26. F. L. Hause, P. L. Middleton, and H. L. Stover, COP X-ray resist technology, *Kodak Microelectron. Sem. Proc., San Diego, California* (October 1979).
27. M. Doken and I. M. Yata, *J. Electrochem. Soc.* **123**, 2235 (1976).
28. L. M. Ephrath, *J. Electrochem. Soc.* **126**, 1419 (1979).
29. K. Tokunga and D. W. Hess, *J. Electrochem. Soc.* **127**, 928 (1980).
30. G. Schwartz, L. B. Zielinski, and T. S. Shopen, *Proc. Symp. Etch. Pattern Definit.* (H. E. Hughes and M. J. Rand, eds.), p. 122. Electrochemical Society, Princeton, New Jersey, 1976.
31. J. A. Bondur, *J. Vac. Sci. Technol.* **1**, 1023 (1976).
32. N. Hosokawa, R. Matsuzuki, and T. Asamaki, *Jpn. J. Appl. Phys. Suppl.* **2**, 435 (1974).
33. H. R. Koenig and L. I. Maissel, *IBM J. Res. Dev.* **14**, 168 (1970).
34. R. A. Heinecke, *Solid State Electron.* **18**, 1146 (1975).
35. R. A. Heinecke, *Solid State Electron.* **19**, 1039 (1976).
36. C. J. Mogab and W. K. Harshbarger, *Electronics* **51**, 117 (1978).
37. C. M. Melliar-Smith, *J. Vac. Sci. Technol.* **13**, 1008 (1976).
38. S. Somekh, *J. Vac. Sci. Technol.* **13**, 1003 (1976).
39. H. I. Smith, *Proc. Symp. Etch. Pattern Definit.* (H. E. Hughes and M. J. Rand, eds.), p. 133. Electrochemical Society, Princeton, New Jersey, 1976.
40. H. W. Lehmann, L. Kraushbauer, and R. Widmer, *J. Vac. Sci. Technol.* **14**, 281 (1977).
41. R. M. Jecmen, C. H. Hui, A. V. Ebel, V. Kynett, and R. J. Smith, *Electronics* **52**, 124 (1979).
42. S. S. Liu, R. J. Smith, R. D. Pashley, J. Shappir, C. H. Fu, and K. R. Kokonen, *IEEE Int. Electron Devices Meeting Tech. Digest, Washington, D.C.* p. 352 (1979).
43. C. M. Osburn and D. W. Ormond, *J. Electrochem. Soc.* **119**, 591 (1972).
44. D. R. Young and C. M. Osburn, *J. Electrochem. Soc.* **120**, 1578 (1973).
45. J. R. Ligenza and W. G. Spitzer, *J. Phys. Chem. Solids* **14**, 131 (1960).
46. N. Tsubouchi, H. Mlyoshi, H. Abe, and T. Enomoto, *IEEE Trans. Electron Devices* **ED-26**, 618 (1979).
47. R. J. Zeto, C. D. Bosco, E. Hryckowian, and L. L. Wilcox, *Electrochem. Soc. Spring Meeting*, Abstr. No. 82, Philadelphia (1977).
48. M. Maeda, H. Kanunoka, H. Shemoda, and M. Takagi, *Symp. Semicond. IC Technol., 13th Tokyo* (1977).
49. W. Kern and V. S. Ban, Chemical vapor deposition of inorganic thin Films, *in "Thin Film Processes"* J. L. Vosson and W. Kern, eds.) pp. 257–331 Academic Press, New York, 1978.

50. W. Kern and G. L. Schnable, *IEEE Trans. Electron Devices* **ED-26,** 647 (1979).
51. W. A. Brown and T. I. Kamins, *Solid-State Technol.* **22,** 51 (1979).
52. J. S. Sandhu and J. L. Reuter, *IBM J. Res. Dev.* **15,** 464 (1970); R. Gereth, A. Rostka, and K. Kreuzer, *J. Electrochem. Soc.* **120,** (1973).
53. Arsenosilicafilm for N-Channel MOS. Emulsitone Co., Whippany, New Jersy.
54. T. Ohzone, T. Hirao, K. Tsuji, S. Horiuchi, and S. Takayanagi, *IEEE J. Solid-State Circuits* **SC-15,** 201 (1980).
55. Y. Wada, S. Nishimatsu, and K. Sato, *Solid-State Electron.* **21,** 513 (1978).
56. S. Matsue *et al.,* A 256K dynamic RAM, *Proc. Int. Solid-State Circuits Conf.* p. 232 (1988).
57. P. D. Parry, *J. Vac. Sci. Technol.* **15,** 111 (1978).
58. P. D. Parry, *J. Vac. Sci. Technol.* **13,** 622 (1976).
59. G. Ryding, *Proc. Int. Microelectron./Semicond. Conf., Anaheim, California* p. 107 (1974).
60. G. Ryding and W. R. Bottoms, High current systems, *in* "Radiation Effects." Gordon and Breach, New York, 1979.
61. M. Y. Tsai, B. G. Streetman, R. J. Blattner, and C. A. Evans Jr., *J. Electrochem. Soc.* **126,** 98 (1979).
62. J. Orton, private communication, 1980.
63. Y. Wada and N. Hashimoto, *J. Electrochem. Soc.* **127,** 461 (1980).
64. M. Nakano, and R. Toger, I. C. Division, Fujitsu Ltd., unpublished manuscript.
65. T. Hirao *et al., J. Appl. Phys.* **51,** 262 (1980).
66. C. W. Nelson, Metallization and glassing of silicon integrated circuits, *Hybrid Microelectron. Symp., Dallas, Texas* p. 413 (1969).
67. W. E. Engler and D. M. Brown, *IEEE Trans. Electron Devices* **ED-19,** 54 (1972).
67a. P. Shah, private communication, 1980.
68. H. Ishikawa, M. Yamamoto, T. Nakamura, and N. Toyokura, A Mo gate 4K static MOS RAM, *Int. Electron Device Meeting Digest Tech. Papers, Washington, D.C.* p. 358 (1979).
69. F. Yanagawa, K. Kiuchi, T. Hosaga, T. Tsachiya, T. Amazawa, and T. Mano, 1 m Mo-gate 64K bit MOS RAM, *Int. Electron Device Meeting Digest Tech. Papers, Washington, D.C.* p. 362 (1979).
69a. M, Kondo *et al., IEEE J. Solid State Circuits* **SC-13,** 611 (1978).
70. T. P. Chow, A. J. Steckl, M. E. Motamedi, and D. M. Brown, MoSi$_2$-gate MOS FETS for VLSI, *Int. Electron Device Meeting Digest Tech. Papers, Washington, D.C.* p. 458 (1979).
71. T. Mochizuki, K. Shibata, T. Inoue, and K. Ohuchi, *Jpn. J. Appl. Phys. Suppl. 17-1* **17,** 37 (1977).
72. K. C. Saraswat, F. Mohammadi, and J. D. Meindl, Wsi$_2$ gate MOS devices, *Int. Electron Device Meeting Digest Tech Papers* p. 462. Washington, D.C., 1979.
72a. S. P. Murarka, Refractory Silicides for Low Resistivity Gate and Interconnects, *IEEE J. Solid State Circuits* **SC-13,** 454 (1979).
73. B. L. Crowder and S. Zirinskij, *IEEE Trans. Electron Devices* **ED-26,** 369 (1979).
74. J. L. Vossen, *J. Vac. Sci. Technol.* **8,** 751 (1972).
75. T. Mochizuki, *Jpn. J. Appl. Phys.* **17,** 37 (1978).
76. J. A. Cunningham, C. R. Fuller, and C. T. Haywood, *IEEE Trans. Reliability* **R-19,** 182 (1970).
77. T. R. Lawson, *Solid State Technol.* **5,** 22 (1966).
78. B. T. Murphy, *Proc. IEEE* **52,** 1537 (1964).
79. W. G. Ansley, *IEEE Trans. Electron Devices* **ED-15,** 405 (1968).

80. R. B. Seeds, *IEEE Int. Electron Devices Meeting* p. 12. Washington, D.C., 1967.
81. A. C. Ipri and J. C. Sarace, *RCA Rev.* **38**, 323 (1977).
82. J. Bernard, *IEEE Trans. Electron Devices* **ED-25**, 939 (1978).
83. J. H. Lee, A. V. S. Satya, A. K. Ghatalia, and D. R. Thomas, Electrical Defeat Monitor Structure, U.S. Patent 3, 983, 479 (1976).
84. R. P. Cenker, D. G. Clemons, W. R. Huber, J. B. Petrizzi, F. J. Procyk, and G. M. Trout, *IEEE Trans. Electron Devices* **ED-26**, 853 (1979).
85. J. W. Lathrop, R. S. Clark, J. E. Hull, and R. M. Jennings, *Proc. IEEE* **55**, 1988 (1967).
86. R. Aubusson and C. Cutt, *IEEE J. Solid-State Circuits* **SC-13**, 339 (1978).
87. S. E. Schuster, *IEEE J. Solid-State Circuits* **SC-13**, 689 (1978).
88. V. G. McKerney, A 5V 64K EPROM utilizing redundant circuitry, *Int. Solid-State Circuits Conf. Digest Tech. Papers* 146 (1980).
89. T. Mano, K. Takeya, T. Wanatabe, K. Kiuchi, T. Ogawa, and K. Hirata, A 256K RAM fabricated with molybdenum-polysilicon technology, *Int. Solid-State Circuits Conf. Digest Tech. Papers* 234 (1980).
90. D. Nyyssonen, *Semicond. Int.* **3**, 39 (1980).
91. P. S. Burggraaf, *Semicond. Int.* **3**, 33 (1980).
92. D. S. Perloff, T. F. Hasan, and E. R. Blome, *Solid-State Technol.* **23**, 81 (1980).
93. I. J. Stemp, K. H. Nicholas, and H. E. Brockman, *IEEE Trans. Electron Devices* **ED-26**, 729 (1979).
94. J. M. Hu, *J. Appl. Phys.* **40**, 4413 (1969).
95. K. G. Moerschel, C. W. Pearce, and R. E. Reusser, *in* "Semiconductor Silicon 1977" (H. R. Huff and E. Sirtl, eds.), p. 170, ECS, Princeton, New Jersey, 1977.
96. B. Leroy and C. Plougonven, *J. Electrochem. Soc.* **127**, 961 (1980).
97. E. W. Hearn, E. H. Tekeat, and G. H. Schwuttke, *Microelectron. Reliability* **15**, 61 (1976).
98. R. E. Gegenwarth and F. P. Laming, Effect of plastic deformation of silicon wafers on overlay, *Proc. SPIE* **100**, 66 (1977).
99. A. S. Grove, "Physics and Technology of Semiconductor Devices." Wiley, New York, 1967.
100. A. Goetzberger and W. Shockley, *J. Appl. Phys.* **31**, 1821 (1960).
101. J. E. Lawrence, *J. Electrochem. Soc.* **112**, 796 (1965).
102. T. M. Buck, T. M. Poate, and K. A. Pickar, *Surface Sci.* **35**, 362 (1973).
103. R. L. Meek and C. F. Gibbon, *J. Electrochem. Soc.* **121**, 444 (1974).
104. T. E. Seidel, R. L. Meek, and A. G. Cullis, *J. Appl. Phys.* **46**, 600 (1975).
105. L. E. Katz, *J. Electrochem. Soc.* **121**, 969 (1974).
106. R. L. Meek, T. E. Seidel, and A. G. Cullis, *J. Electrochem. Soc.* **122**, 789 (1975).
107. T. Y. Tan, E. E. Gardner, and W. K. Tice, *Appl. Phys. Lett.* **30**, 175 (1977).
108. G. A. Rozgonyi and C. W. Pearce, *Appl. Phys. Lett.* **31**, 343 (1977).
109. S. P. Murarka, *J. Electrochem. Soc.* **123**, 765 (1976).
110. G. A. Rozgonyi, P. M. Petroff, and M. H. Read, *J. Electrochem. Soc.* **122**, 1725 (1975).
111. H. R. Huff, Silicon materials for VLSI/VHSIC, *in* "Integrated Circuit Technologies for the 1980's." Univ. of California Continuing Education in Engineering, Palo Alto, California (Feb. 12, 1980).
112. P. M. Petroff, G. A. Rozgonyi, and T. T. Shing, *J. Electrochem. Soc.* **123**, 565 (1976).
113. H. Shiraki, *Jpn. J. Appl. Phys.* **15**, 1 (1976).
114. L. E. Katz and D. W, Hill, *J. Electrochem. Soc.* **125**, 1151 (1978).
115. W. A. Keenan and R. O. Schwenker, Gettering of bipolar devices with diffused isola-

tion regions, *in "Semiconductor Silicon"* (H. R. Huff and R. R. Burgess, eds.), p. 701. ECS, Princeton, New Jersey, 1973.

116. J. G. Skinner, Life with EBES, *in Kodak Microelectron. Seminar* (1976).
117. M. Dan, I. Tanabe, M. Hoga, and S. Torisawa, Fabrication system of high quality hard-surface masks for LSI's, *in Kodak Microelectron. Seminar* (1979).
118. K. A. Snow, Optical problems of small geometry automatic mask inspection, *Proc. SPIE* **135,** 96 (1978).
119. J. G. Skinner, Use of an automated mask inspection system (AMIS) in photomask fabrication, *Proc. SPIE* **100,** 20 (1977).
120. P. Sandland, Automatic inspection of mask defects, *Proc. SPIE* **100,** 26 (1977).
121. Dataquest Research Newsletter **1,** 3 (October 11, 1978).
121a. W. Kauffman, private communication, 1980.
122. N. Wu, Processing and device technology advancements, *Proc. Int. Solid-State Circuits Conf.* 208 (1980).
123. Texas Instruments Incorporated, Dallas, Texas, Annual Report, p. 8 (1979).

Chapter **2**

Principles of Optical Lithography

MICHAEL C. KING

Microlithography Division
The Perkin-Elmer Corporation
Norwalk, Connecticut

41

LIST OF SYMBOLS

A	Amplitude
a_i, b_i	Fourier coefficients
D^2	Reduction in MTF caused by one Rayleigh unit of defocus
$E(x)$	Exposure at position x
E_M	Exposure in large clear areas
E_0	Photoresist threshold exposure
f	f/number
h	Resist removed by anisotropic etch
I	Intensity of illumination
L	Width of the resist profile at the resist–substrate interface
M	Modulation
M_I	Image modulation with incoherent illumination
M_P	Image modulation with partial coherence
MTF	Modulation transfer function (incoherent illumination)
N	Numerical aperture
p	Period of mask pattern
Q	Ratio of operating point exposure with partial coherence to operating point exposure with incoherent illumination
R	Defocus in Rayleigh units
S	Degree of partial coherence
$T(E)$	Resist thickness after development due to an exposure E
T_0	Nominal resist thickness before development
u	Horizontal resist loss produced by anisotropic etch
x	Positional coordinate
y	Mask-to-wafer separation in a proximity printer
z	Distance from plane of best focus
α	Absorption coefficient to the base e
γ	Photoresist gamma
Δ	Variation in the variable that follows
δ	Depth of focus
θ	Resist profile angle
ν	Spatial frequency $(= 1/p)$
Φ	Angle of diffracted light
ϕ	Phase angle with coherent imagery
Ψ	Half-angle subtended by optical aperture
Ω	Half-angle subtended by illumination

To distinguish among various forms of illumination we shall use the following subscripts:

C	Coherent
I	Incoherent
P	Partial coherence

I. INTRODUCTION

A. Generalized Photolithographic Systems

During the process of microfabrication, fine-line stencil masks are formed using thin photosensitive polymers, called *photoresists* or *resists,* which selectively protect an underlying wafer substrate against chemical or physical attack. Optical lithography, as the name implies, refers to a means for patterning the resist using actinic radiation in the optical region of the electromagnetic spectrum.

A generalized photolithographic system is depicted in Fig. 1. The information to be replicated is delineated on a thin optically opaque layer supported by a transparent substrate. This pattern, called a *mask,* is transferred or imaged by a lithographic exposure system to form an aerial image, which consists of a spatially dependent light intensity pattern in the vicinity of the wafer. By exposing the resist-coated wafer to the aerial

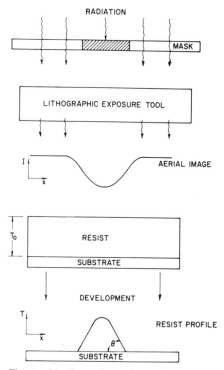

Fig. 1.   Idealized photolithographic system.

image, the resist becomes more soluble (with a positive photoresist) to a chemical developer. This allows for easy removal of the exposed regions. The resist image remaining on the wafer behaves as a stencil for subsequent processing such as etching or metallization.

A perfect lithographic process is achieved when the aerial image duplicates the mask pattern without loss of information or distortion. Similarly, the perfect resist process responds with full resist thickness for an exposure intensity below a threshold value and with complete removal of the resist for exposures above the threshold. It is in the nature of the physical and chemical processes that a perfect lithographic system is never realized. As the minimum geometries on the mask get smaller, the lithographic system behaves less ideally. This has stimulated an ever-increasing effort in lithographic technology to keep pace with the ever-shrinking geometries associated with VLSI fabrication.

B. Historical Perspective

Modern-day optical lithography is a hybridization of very new and very old technologies. Most of the techniques in use today were perfected over the past 20 years in support of semiconductor device fabrication. However, the development of photoresist processes can be traced back nearly 150 years to the pioneering photoengraving work of Niepce [1]. The earliest work in microphotography goes back nearly as far to the accomplishments of Dancer [2] in 1839. With the advent of solid-state integrated circuitry (IC) in the early 1960s, optical lithography became established as the primary means for device fabrication. Using photoresists developed in the 1950s for thin-film and printed circuit board applications, contact printing was readily adapted to satisfy the needs of the emerging industry.

Appreciating the status of today's optical lithographic technologies requires an understanding of how attitudes toward lithographic characterization have changed over the past 10 years. As late as 1970, it was customary to evaluate lithographic exposure tools by the term "resolution"; that is, "this machine is a 10-μm machine," or "that one does 5 μm." Resolution meant the smallest achievable linewidth under ideal exposure conditions. In time, IC industry engineers began to realize that minimum resolution did not relate to performance under manufacturing conditions. In response, they began referring to the "minimum feature size," or "minimum working feature." This referred to the smallest feature with which devices could be fabricated with satisfactory yield.

By 1975 devices were being fabricated with features smaller than 4.0 μm. At this level a single descriptive parameter proved insufficient to describe lithographic performance. It was often found that the exposure technology that provided the highest resolution with acceptable yield did not necessarily provide the best performance for larger features. At about the same time, requirements for linewidth control and overlay accuracy began to exceed the performance provided by some of the available lithographic exposure equipment. At this juncture, the expression "minimum feature size of x micrometers" took on the added meaning that the exposure tool provided all the characteristics required by an x-micrometer design rule.

The single parameter, "minimum feature size," with its implied meaning, was an adequate description of lithographic exposure tools until the advent of 2.0-μm design rules in 1979. At this level it became impossible to judge lithographic performance without considering nonhardware aspects of the technology. With overlay requirements in the submicrometer region and linewidth control tolerances under 0.2 μm, the role of the photoresist and subsequent processes could not be ignored. It is partially because of these factors that different IC manufacturers with identical exposure equipment, manufacturing identical types of devices (NMOS, for example), achieved minimum features ranging from as small as 1.5 μm to as large as 3.0 μm.

Characterization of today's optical lithographic performance and predictions for future advancements require a three-part evaluation. First, we must investigate the nature of the optical imagery independent of the effects of resist and process. This is referred to as the *aerial image*. Second, we must investigate the nature of the resist image. By this we mean the resist profile after development. The aerial image I and the resist profile T are depicted in Fig. 1 of the generalized exposure system with positive resist. Third, we must investigate the consistency of the imagery over the entire wafer. This includes the variations in linewidth with position and the accuracy to which the features can be placed over the wafer surface. As in the case of minimum feature size, both characteristics are dependent on the process as well as the exposure tool.

II. OPTICAL EXPOSURE TECHNIQUES

A. Contact/Proximity Printing

Contact printing, as the name implies, is simply bringing the resist-coated wafer in physical contact with the mask and illuminating the back of the mask with actinic radiation. The wafer is held flat against a vacuum chuck and the back of the chuck swivels in a spherical socket. Contact is achieved by vertically transporting the wafer chuck assembly until the wafer is in intimate contact with the mask surface. The socket allows the chuck to tilt in order to achieve planarity with the mask. To align the mask pattern to the wafer, a high-powered microscope objective (actually a pair of objectives connected to a split-field microscope) is brought behind the mask, allowing the operator to view the mask and wafer patterns simultaneously. With the microscope in position, the wafer chuck assembly is withdrawn to provide an air gap between the mask and wafer surfaces. Alignment is achieved by translating or rotating the wafer until its pattern overlays the mask pattern. When the wafer is brought back into contact, the microscope is moved away and an exposing source of illumination is brought into position. Exposure is achieved using a highly collimated source of ultraviolet light originating from a high-pressure mercury arc lamp.

Contact printing refers to the situation when the mask and wafer are in intimate contact during exposure. In proximity printing, exposure takes place with a finite gap between the substrates. In practice, neither condition is completely achieved. In its chucked state the wafer does not assume a flat surface. This limits the total wafer area that can achieve perfect contact. In addition, the resist gives off nitrogen gas during exposure, which further separates the mask and wafer. During proximity printing, the mask and wafer are separated by about 10 μm during exposure. This is done to reduce mask and wafer damage that results during contact. Unfortunately, the same wafer surface irregularities that prevent perfect contact also inhibit a clean separation, and damage still results. This situation is also aggravated by the need for some physical contact to achieve mask-to-wafer parallelism.

The advantages inherent in contact/proximity printing are the simplicity of the system and the relatively small capital investment. These techniques are capable of producing good 1.0-μm features in a "contact" mode and approximately 4.0 μm in a "proximity" mode. Because they incorporate a minimal number of optical elements and broadband illumination, they are capable of very high exposure rates.

There are many difficulties that limit the usefulness of contact/proximity printing in VLSI applications. Induced mask defects limit the mask life from ten to a few hundred exposures depending on the severity of the contact and the size of each die. The economics of using a disposable mask dictate the use of an inexpensive mask. This is usually a photographic emulsion, which imparts an additional limitation on practical feature size and overall yield. The process of clamping the wafer between two substrates also distorts the wafer surface in an unpredictable and nonrepeatable fashion. Overlay errors as large as 1.0 μm are easily produced. To limit penumbra effects, contact printers use a highly collimated source of illumination. The coherency tends to accentuate diffraction effects in the imagery. A practical solution is to decollimate the illumination by a few degrees with a subsequent loss in minimum feature size.

Proximity printers operate in the region of Fresnel diffraction where the smallest achievable feature is proportional to $(\lambda y)^{1/2}$, where λ is the exposing wavelength and y the mask-to-wafer separation. By decreasing the wavelength, y can be increased without loss in resolution. In recent years there has been a renewed interest in proximity printing using deep UV (2000 to 2600 Å) illumination, since reducing the wavelength from 4000 to 2000 Å and doubling the mask-to-wafer separation provide a marked improvement in yield. Another benefit of proximity printing is that it provides considerable improvement in linewidth control compared to projection printing (this will be explained in Section V.B). By decreasing the wavelength, the improvement is even greater.

B. Projection Printing

1. Full-Field Exposure

The desire to avoid the mask damage induced by contact printing led to many early attempts to *project* the mask image onto the wafer. At a time when the largest wafer was 2 in. in diameter, equipment was available that could expose the entire wafer to an image of unity magnification with a single exposure. These tools were capable of exposing minimum features down to 5.0 μm with optical distortions of about 2.0 μm. However, they never proliferated because of the strong competition from proximity printing. With a trend toward smaller features and larger wafers, the pursuit of full-field exposure systems was dropped. The optical systems did not easily scale to the larger field size without sacrifice in image distortion. In addition, the requirement for uniform illumination over the larger field proved to be a major problem.

Another early attempt to get away from contact printing saw the development of 4× and 10× reduction refractive systems that imaged a subfield of the wafer. These systems were capable of features down to 1.0 μm and were, at the time, useful for a few specialized applications. However, because the wafer stage required manual realignment before each exposure, the throughput was impractically slow and the equipment never became popular.

2. Ring Field Scanning

In 1973 a new concept in projection printing utilizing a "ring field" optical system was introduced [3]. As shown in Fig. 2, the system consists of a primary and a secondary mirror having centers of curvature located on a common axis. Centered on this axis is a ring-shaped field over which a high degree of optical correction is obtained. Each point on the ring is imaged to its diametrically opposed position. The ring has an effective width of approximately 1.0 mm. The ring field system is adapted to a projection printer by folding the light path with three flat mirrors, as shown in Fig. 3. In this manner, an arc 1.0 mm wide and approximately 3 in. long is imaged from the mask to the wafer. Full-field exposure is obtained by mechanically scanning the mask and wafer through the ring field position.

The ring field scanning system offers many advantages over the full-field, single-exposure approach. The all-reflective optical system imposes no bandwidth restrictions on the actinic wavelengths. It is also easier to control light scattering and temperature effects with reflective optics. The telecentricity inherent in the ring field design maintains 1:1 magnification independent of focus variations. Exposing the wafer section by section also reduces the amount of stray and scattered light reaching the wafer. This is most important when using negative resists.

Today, ring field 1:1 mask aligners have become an industry standard.

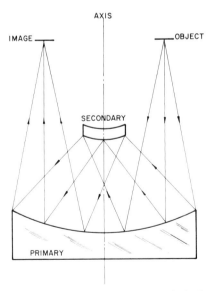

Fig. 2. Ring field optical system. Two concentric spherical mirrors image a narrow annular field on one side of the axis to a similar field on the other side.

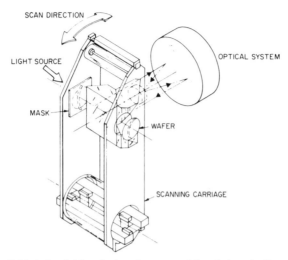

Fig. 3.   Folded ring field optical system as used in a 1:1 projection printer.

Using an $f/3$ optical system exposing in a 3400- to 4400-Å bandwidth, they are capable of lithography at a 2.0-μm design rule with throughput rates approaching sixty 4-in. wafers/hr. It has been demonstrated that by redesigning the optical coatings for maximum efficiency at 3000 Å and using an actinic bandwidth between 2800 and 3200 Å, it is possible to achieve a performance consistent with 1.5-μm design rules while using the same $f/3$ optical system.

3. Small-Field, Step-and-Repeat System

In the late 1970s, a number of small-field imaging systems that could image a small number of die at a time became available. Unlike their predecessors, the wafer is moved between exposures on an interferometrically controlled $x-y-z$ stage. In this manner, the wafer can, in principle, be accurately positioned and focused for each subfield exposure. The most common configuration incorporates a $10\times$ refractive reduction lens corrected at 4040 or 4350 Å with a 14-mm diameter exposure field. The step-and-repeat exposure systems incorporate a relatively fast optical system operating between $f/1.5$ and $f/2.5$. This results in a minimum working feature in the 1.0- to 2.0-μm region. The inherently small depth of focus can be accommodated by automatic focusing systems.

When considering the extension of optical projection lithography to submicrometer imagery, the advantages of a step-and-repeat protocol are apparent. With the potential to align every separate die, it is possible to achieve excellent level-to-level overlay in the presence of process-induced wafer distortion. Another major advantage is that a 10× reduction system uses a reticle at 10 times the magnification of the mask used in a 1:1 system. This eliminates most of the problems associated with fabricating and inspecting a mask with submicrometer features. On the other hand, the difficulty of making a perfect reticle and keeping it free of printable defects poses a considerable challenge.

In the 1.0- to 2.0-μm region both 1:1 scanning and step-and-repeat systems involve different tradeoffs. The step-and-repeat system has the advantage that high-quality reticles are easier to fabricate and inspect than 1:1 masks. With die-by-die alignment, steppers have a potential for tighter overlay tolerances. Scanning systems have the advantage of better cost effectiveness and considerably faster throughput rates while maintaining comparable imagery when used at 3000 Å. In the next few years it is very possible that we will see a hybridization of scanning along with step-and-repeat systems on the same production lines to take maximum advantage of their different characteristics. It is possible that future developments combining deep UV exposure with reflective step-and-repeat optical systems will be capable of extending optical lithographic performance to the 0.5-μm level.

III. PHOTORESISTS AND SUBSTRATES

A. General Properties of Photoresists

The properties of photoresists are as important as the properties of the exposure tool in achieving good lithographic performance. Photoresists can be classified by primary and secondary characteristics as listed in Table I. The primary characteristics relate directly to the quality of the imagery. The secondary characteristics describe the properties required to maintain acceptable yield in a manufacturing environment. In this section we shall discuss the primary characteristics. In Section IV. C we shall show how these properties affect lithographic performance.

Resist systems can be divided into the broad classes of positive and negative resists, depending on how they respond to the actinic radiation. In a positive resist system, the exposed areas become more soluble than the unexposed regions. Most commercial positive resists contain a photosensitive dissolution inhibitor, an alkaline-soluble base polymer (usually a novolac resin), and a solvent. The dissolution inhibitor prevents the base polymer from dissolving in the alkaline developer. After exposure to light, the dissolution inhibitor concentration is reduced in selected areas, leading to an increased solubility.

TABLE I

Resist Characterization

Primary characteristics	Secondary characteristics
γ	Swelling
Absorption	Adhesion
Threshold E_0	Dry and wet etch resistance
	Process latitude
	Ease of application
	Uniformity of coating
	Thermal stability
	Bottle life
	Environmental concerns
	Contamination

A second type of positive resist action takes place when a positive electron or x-ray resist is exposed to deep UV radiation. In this situation, the radiation has enough energy to cause main-chain fissions in the base polymer. The lower molecular weight in the exposed regions leads to greater solubility during development. Some typical examples are such materials as PMMA and PMIPK.

Negative resists become less soluble in the regions exposed to light. The mechanism that leads to this decrease in solubility is the cross-linking of the base polymer, which increases the molecular weight in the exposed areas. In the early days of microlithography, negative resists were used exclusively. They offered a high degree of process latitude and provided excellent secondary properties. Unfortunately, there are characteristics of negative resists that limit their usefulness. First, in the process of development, both the exposed and unexposed areas are permeated by the developer, which produces a swelling in the resist profile. This action limits the useful feature size to about 2.0 μm. Second, when using a projection printer, there is always a finite level of background light that originates from light scattered off the projection optics. With positive resists the background light simply reduces the height of the resist profile and, consequently, has little effect on the overall quality of the imagery. With a negative resist, the effect is more serious. The background light crosslinks a thin film in the top surface of the resist. This film usually becomes punctured and falls down between the features and results in a "scumming" effect. Experiments conducted to measure the effect of background light on scumming have concluded that as little as 1% scattered light can ruin the imagery of 1.0-μm lines and spaces. Third, standing-wave phenomena (see Section III. C) can produce an inadequate exposure at the resist–wafer interface. If this occurs and the features are small, the developed pattern will often wash away.

B. Optical Properties of Positive Photoresists

When predicting how a specific resist will respond to a given aerial image, it is important to know how the light is absorbed as it passes through the resist film. It is useful to measure the percentage of light transmitted by the resist as a function of wavelength. A typical example is given in Fig. 4. By noting the resist thickness T_0, the coefficient of absorption α can be calculated from the transmission curve

$$\alpha = T_0^{-1} \log_e(\text{fraction transmitted}).\tag{1}$$

In general, it is desirable to have no less than 60% transmission at the actinic wavelengths. When the percentage transmission falls below this value, the exposure gradient through the resist depth degrades the image profile, since a large percentage of the light will be absorbed by the upper

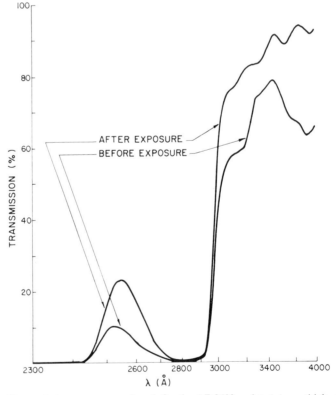

Fig. 4. Transmission versus wavelength for the AZ-2400 resist, 1.1 μm thick, measured before and after exposure.

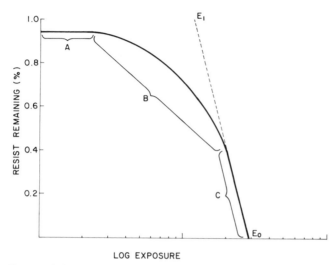

Fig. 5. Characteristic curve for a positive resist. Threshold exposure is E_0. The photoresist γ is defined by drawing the tangent at E_0. The log scale is to the base e.

layers of resist. If the percent transmission is too great, the resist sensitivity will drop, resulting in impractical exposure times. Percent transmission or absorption curves are valuable for predicting the performance of resists at specified wavelengths. To some degree, light is absorbed by all of the constituents in the photoresist. The change in absorption with exposure is caused by the reduction in photoactive components that occurs during exposure. It is tempting to associate sensitivity with absorption, but this is not necessarily the case. The high absorption below 3000 Å that is found with most positive resists is a characteristic of the base polymer and not the photosensitized elements.

A most useful characterization of a photoresist is obtained by measuring the percentage of resist remaining after development as a function of exposure. By plotting this "characteristic curve" on a semilog scale, we can obtain information about the resist that allows us to predict the quality and consistency of the lithographic imagery. Figure 5 shows the basic features of a characteristic curve for a typical positive resist. In region A there is a range of low-level exposures for which the solubility of the resist is unmodified. A loss of approximately 10% of the resist thickness in this region demonstrates that the resist has a finite solubility even when no exposure takes place. In region B the solubility begins to increase with exposure. In region C the curves become linear, with log E indicating that light is being absorbed within the resist layer. It is customary to identify two parameters, E_0 and γ, with the characteristic curve: E_0, the

TABLE II

Photoresist γ^a

Exposing bandwidth (Å)	Resist	γ
3400–4400	AZ-111	1.3
	AZ-1370	2.0
	AZ-2400	2.0
	HR-204	1.43
3000–3300	AZ-2400	1.67
	Kodak 809	1.61
	AZ-1370	1.2
	HR-204	1.0
2800–3300	AZ-2400	0.96
	AZ-1370	1.2
	HR-204	0.85

[a] Measured values of γ (as defined in text) for $f/3$ reflective projection printer. Substrate is silicon wafer, coated with 7000 Å of wet oxide. Resist thickness is 1.0 μm with standard processing.

minimum exposure needed to remove completely the photoresist during development, is called the "threshold exposure" and is a measure of the sensitivity of the process; γ, the slope of the characteristic curve as measured in region C, is defined by the expression

$$\gamma = [\log_e(E_0/E_1)]^{-1}, \tag{2}$$

where E_1 is the exposure corresponding to the intersection of the tangent drawn at E_0 and the ordinate at 100% resist thickness before development. This definition of γ and E_0 differs from previous definitions in three important ways. First, in previous treatments γ has been defined by drawing the tangent at the point on the characteristic curve corresponding to 50% thickness. In this chapter we have drawn the tangent at the point where the resist thickness goes to zero. This has been done since our main interest is in the behavior of the resist profile near the resist substrate interface. Second, γ has been defined to the base e. This has been done to simplify many of the equations that follow. Third, in the past E_0 and γ were considered to be solely functions of the resist type and the processing conditions. For our purposes, this description is too restrictive because it overlooks the fact that E_0 and γ are also functions of resist thickness, the optical properties of the underlying substrate, and the spectral composition of the actinic radiation. It is the primary purpose of this chapter to describe the basic functional relationships that connect the resist process and exposure parameters with lithographic performance. For this reason it is necessary to expand the meanings of E_0 and γ to include all the parameters on which they are functionally dependent. To achieve this, we will redefine E_0 and γ to mean the exposure and slope at threshold as determined from a characteristic curve as measured with a specific exposure tool, development process, resist thickness, and substrate composition. This expanded definition is consistent with a definition first proposed by McGillis and Fehrs [4]. Table II lists some representative values of γ measured on 7000 Å of oxide on silicon wafers.

C. The Wafer Substrate

The material composition of the substrate can have a dramatic effect on lithographic imagery. Depending on the optical properties of the substrate, a large percentage of the light can be reflected back into the photoresist. Some of this light becomes trapped in the resist (or oxide) film with each of the repeated passes back and forth between the substrate and the resist–air interface until it is completely absorbed. In the resist there will be regions where the amplitudes of light interfere destructively. These areas are called *nodes* and correspond to regions of minimum exposure. Between the nodes are regions of constructive interference called *antinodes* where the exposure is maximum. The distance between nodes is $d = \lambda/2n$, where n is the optical index of the photoresist. The resist threshold E_0 is a function of the position of the node closest to the resist substrate boundary. When a node falls on the interface, the threshold exposure is increased. Likewise, an antinode at the interface lowers the threshold. Such standing waves impart a fine structure to the characteristic curve with a periodicity of $\lambda/2n$ in depth. The wavelength of the actinic radiation and the thickness and optical properties of the underlying substrate determine the placement of the nodes within the substrate, whereas the thickness of the resist affects the intensity of illumination at the node and antinode positions. When the various parameters that describe the resist and substrate are known, it is possible to predict the exact effect of standing waves on a specified exposure condition [5,6]. It is common practice to choose the thickness of the thin films and the development conditions to minimize the effects of standing waves for a given substrate and exposing wavelength, but a major problem occurs if we wish to control the feature linewidth when the substrate has variable optical properties and film thicknesses.

Consistent with our approach, it is sufficient that we measure the characteristic curve for each substrate condition that will occur during device fabrication. In Section V we shall show how this information can be used to estimate linewidth control.

D. Substrate Topography

During the process of fabricating devices, patterns are etched into the substrate. In addition to modifying the standing-wave phenomena in subsequent exposures, the substrate topography affects the thickness of the photoresist, which in turn affects the characteristic curve. At the top of the topographical steps the resist is thinner, whereas at the bottom it is thicker.

The difficulty of achieving linewidth control when imaging over steps can be minimized by optimized selection of the resist–actinic radiation combination. Using resists with the least absorption reduces the functional dependence of γ and E_0 on resist thickness. Solutions to this problem are discussed in Sections III. E and V. D.

E. Multilayer Resist Technology

During the past few years, a new process called *multilayer resist technology* [7,8] has been developed. The process incorporates either two or three levels to achieve similar results. In the three-level system the wafer is first coated with a thick layer of resist. Onto this layer is deposited a thin substrate, such as chrome, which is, in turn, coated with a thin layer of photoresist whose properties may differ from those of the first layer. The thin top layer is exposed in a lithographic exposure tool and developed, and the pattern is etched into the metal. The lower resist layer is either exposed to UV light or it is ion-milled using the chrome layer as a mask. The net result is a high-resolution resist profile with vertical sidewalls.

With this process the thick lower resist layer fills in the topography to provide a flat surface for imaging. The intermediate metallized level provides uniform standing-wave phenomena for good linewidth control. The thin photo-sensitive layer provides an optimum situation for high-resolution lithography and is insensitive to the optical properties of the wafer substrate. The resolution enhancement achieved with the multilayer process is very dramatic. The process may have more impact on the future of device lithography than some of the novel exposure systems presently under development. For further treatment of this subject, see Chapter 3 in this volume.

IV. THE CHARACTERIZATION OF THE LITHOGRAPHIC IMAGE

A. Modulation Transfer Function

The quality of the aerial image relative to that of the mask pattern is determined by the modulation transfer function (MTF) of the lithographic exposure tool. The significance of the MTF is best elucidated by the following analysis. Consider a mask consisting of lines and spaces of spatial frequency ν, defined as the inverse period ($\nu = 1/p$). Illuminating the mask with light, we can move a linear intensity monitor behind the mask and measure the intensity as a function of position. Noting the maximum and minimum intensities measured behind the mask, we can define the mask modulation for a spatial frequency ν as

$$M_{\text{mask}} = (I_{\text{max}} - I_{\text{min}})/(I_{\text{max}} + I_{\text{min}}). \tag{3}$$

Similarly, by measuring the light intensity as a function of position of the aerial image in the image plane of the lithographic tool, we can define the image modulation by an analogous expression. The MTF of the imaging system is then defined by the expression

$$\text{MTF}(\nu) = M_{\text{image}}(\nu)/M_{\text{mask}}(\nu). \tag{4}$$

This expression is only valid when the mask has a sinusoidally varying transmission. In most photolithographic applications, the mask consists of opaque metal features on a glass substrate. In this case, the mask modulation is 1.0, independent of the feature size. Whereas a sinusoidal pattern diffracts light into one order, a square wave pattern, as found on a photolithographic mask, diffracts light into an infinite number of orders. In high-resolution applications, all orders except the first are cut out by the optical system. In this special case the ratio of the sinusoidal MTF to the image modulation is $\pi/4$. In the treatment that follows, the MTF will always represent the modulation that would occur when imaging a sinusoidal pattern.

From the standpoint of lithographic imagery, MTF provides a reasonably complete characterization of an optical system. Once the MTF is

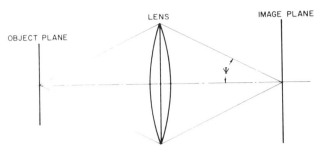

Fig. 6. Simple imaging system. The numerical aperture is $N = n \sin \Psi$. The f/number
$= 1/2N$.

known, accurate predictions can be made for lithographic performance
under a variety of conditions.

Consider the simple imaging system drawn in Fig. 6. The angle Ψ, subtended by the maximum pupil diameter and the image plane, determines
how small a feature can be imaged. This characteristic is described by
either the numerical aperture $N = n \sin \Psi$ or the effective f/number $=$
$1/2N$. In this chapter f/number is represented by the letter f. The variable
n is the index of refraction of the surrounding media, which is air in most
lithographic applications ($n \approx 1$). The nature of the imagery is also
dependent on how the mask is illuminated and the wavelength λ of the
light. In Fig. 7, we show a simplified situation where the mask has a sinusoidally varying transmittance of period p and the illumination is collimated perpendicular to the mask surface. The mask behaves in a manner
similar to a grating by diffracting light into two (± 1) orders. The recom-
bining of the diffracted orders with the undiffracted light (zero order) at
the image plane produces a perfect image of the grating. As the period of
the pattern is decreased or the wavelength is increased, the angle Φ be-

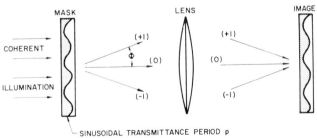

Fig. 7. Imaging a mask with a sinusoidally varying transmittance of period p with coherent illumination. As p is decreased, the diffracted angle Φ becomes larger. An image will
form when $\sin \Phi \leq N$.

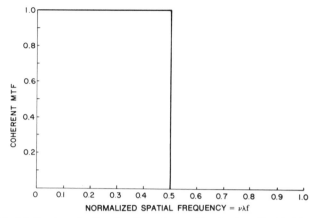

Fig. 8. Modulation transfer function for coherent illumination. The limiting feature size is a grating of period $p = 2\lambda f$.

comes larger. As long as $\Phi \leq \Psi$, all the diffracted light is collected, resulting in a perfect image. Remember that we are considering an ideal optical system free of aberrations. This is a realistic assumption considering the degree of perfection achieved in state-of-the-art lithographic systems. It is easy to imagine how this optical system will treat various size gratings. All gratings should be imaged equally well until the frequency of the grating, $\nu = 1/p$, exceeds $1/2\lambda f$. After this point no image at all will be formed. This is what we would observe if we measured the MTF as a function of the grating frequency. The MTF drawn in Fig. 8 has a value of 1 up to a cutoff frequency $\nu_c = 1/2\lambda f$, after which the modulation goes to zero. Systems that behave in this manner are labeled *coherent*. In practice, the behavior of both contact/proximity printers and projection printers illuminated with coherent collimated light is similar to that of coherent systems.

Another important situation occurs when the illumination subtends a solid cone angle $\Omega > \Psi$. In this case the illumination is said to *overfill* the projection optics and the system is described as *incoherent*. As shown in Fig. 9 with incoherent illumination, light can be diffracted by an angle 2Ψ and still be collected by the projection optics. This leads to the conclusion that incoherent optical systems can image gratings having twice the spatial frequency of the maximum frequency imaged in the coherent case. Figure 10 shows the MTF associated with an incoherent optical system. The MTF appears to decrease monotonically as the frequency is increased. This behavior can be explained from simple geometrical arguments [9].

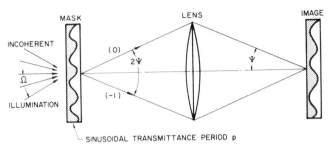

Fig. 9. Imaging a mask with a sinusoidally varying transmittance of period p with incoherent illumination. An image will form when the diffracted angle $\leq 2\psi$. Comparing with Fig. 7, incoherent systems have twice the resolution of that obtained with an equivalent coherent system.

The imagery formed by coherent and incoherent systems differs in appearance. It can be observed that the coherent image has a steeper slope but is degraded by "ringing" at the top of the feature. Coherent systems are linear in amplitude. In other words, the imagery is formed by adding the amplitudes of the various spatial frequency components. The image is thus dependent on the relative phase of each component. Coherent systems are thus degraded by small imperfections in the optical system that vary the phase. In contrast, incoherent systems are linear in intensity and, therefore, less sensitive to phase variations.

For photolithographic applications, it is desirable to compromise between the attributes of coherent and incoherent imagery. This is achieved by partially filling the pupil of the projection optics. The condition thus created is called *partial coherence*. It is characterized by the

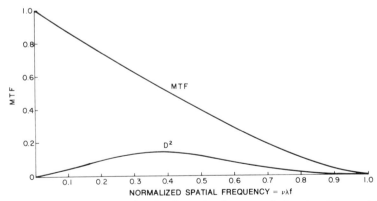

Fig. 10. Modulation transfer function for an incoherent optical system. The modulation is reduced by R^2D^2 when the system is defocused by R Rayleigh units (from Offner [10]).

parameter S, the ratio of the numerical aperture of the illuminator and the numerical aperture of the projection optics. A coherent system is characterized by $S = 0$ and an incoherent system by $S = \infty$, implying that the pupil is overfilled. The difference between $S = \infty$ and $S = 1$ is small.

Analyzing the behavior of a partially coherent imaging system is difficult because the functional relationship between the object and image spectral components is nonlinear. However, it is possible to measure the image modulation for a partially coherent system. It is also possible to calculate it, but, because of this nonlinearity, the concept of MTF is invalid. There is no transfer function from which one can predict the exact imagery that will result from a given mask pattern.

When describing lithographic systems, it has become common practice to use MTF curves to describe the quality of performance even when the system is illuminated in a partially coherent manner. This may trouble the purist, but there is good justification for maintaining the tradition. For fine-line lithographic applications, the exposure tools are used near the limits of performance. In these cases only the fundamental frequency of the mask pattern reaches the image plane. In the sections that follow, we shall demonstrate how many useful characteristics of the lithographic image can be predicted by knowing only the modulation of the fundamental frequency. When the mask features are so large that many spatial frequency components are imaged, the partially coherent MTF ceases to be meaningful. This poses no problem since lithographic performance is nearly perfect for these large features and performance is predictable without resorting to the MTF concept.

Offner [10] has calculated a series of MTF curves based on diffraction considerations as a function of partial coherence S. These curves are plotted in Fig. 11 as a function of the normalized spatial frequency $\nu_n = \nu\lambda f$. The curve for the incoherent MTF for a circular pupil is approximated by the expression

$$M_1(\nu) \cong 1 - (4/\pi) \sin(\nu\lambda f) \qquad (5)$$

for all but the highest spatial frequencies.

The MTF curves in Fig. 11 represent an idealized optical system maintained in perfect focus. Today's lithographic projection optics are derived from sophisticated optical designs and are fabricated to very high tolerances. For this reason, the values obtained from Fig. 11 are in very close agreement with the values measured on available lithographic equipment.

In practice, neither the mask nor the wafer is perfectly flat, so the effects of focus must be noted. By convention, we measure defocus in units of $2\lambda f^2$, referred to as the Rayleigh unit. Offner [10] has calculated the effect of defocus on the incoherent MTF as a function of spatial frequency.

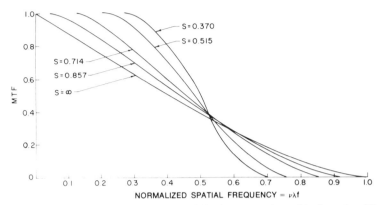

Fig. 11. MTF for imaging system with partial coherence, where S is the ratio of the numerical aperture of the illumination system to that of the projection optics (from Offner [10]).

This is represented by the curve D^2 in Fig. 10. When the optical system is defocused by R Rayleigh units, the MTF can be calculated by subtracting R^2D^2 from the focused MTF value. The depth of focus of an optical system is usually designated as ± 1 Rayleigh unit. However, as we shall see in Section VI, the effective depth of focus for a lithographic system can be smaller or many times larger than the Rayleigh unit depending on the feature being imaged and the resist process employed.

B. Aerial Image

When the mask pattern is an odd periodic function of position with equal lines and spaces of spatial frequency ν, it can be written as an infinite Fourier series

$$\text{mask pattern}(x) = a_0 + \sum_{k=1}^{\infty} a_k \sin(2\pi k\nu x), \qquad (6)$$

where the constants $a_0, a_1, \ldots, a_n$ are the Fourier coefficients of the pattern on the mask. When the optical system is illuminated with incoherent light, the system is linear in intensity. From our definition of MTF in Eq. (4), we see that the Fourier coefficients of the aerial image are simply those of the mask pattern weighted by the MTF of the lithographic tool. Thus,

$$I(x) = a_0 + \sum_{k=1}^{\infty} \text{MTF}(k\nu)a_k \sin(2\pi k\nu x). \qquad (7)$$

In order to simplify the discussion, we shall now state some basic restrictions that will hold throughout the remainder of this chapter. First, we shall only consider the case in which the mask consists of equal lines and spaces of period p ($\nu = 1/p$) extending over the entire mask surface. Second, we shall consider lithographic systems performing near their limiting capability, where only the fundamental spatial frequency of the periodic pattern plays a significant role. In Eq. (7) we shall set $MTF(k\nu) = 0$ for $k > 1$. The preceding restrictions lead to tractable solutions of a simple form, which give clear insight into the performance of lithographic systems.

By applying these two restrictions to Eq. (7) and using the fact that $MTF(0) = 1$, we can derive the following expression for the incoherent aerial image intensity:

$$I(x) = \tfrac{1}{2}I_{max}[1 + M_I(\nu)(4/\pi) \sin(2\pi\nu x)], \tag{8}$$

where I_{max} is the intensity of illumination that would occur in a large, unpatterned area. The term $4/\pi$ is a consequence of the square wave intensity pattern of the mask.

By applying the same restrictions to the case of coherent imaging, we can derive the following equation for the amplitude of the aerial image:

$$A(x,z) = \tfrac{1}{2}A_{max}[1 + (4/\pi)e^{j\phi(z)} \sin(2\pi\nu x)]. \tag{9}$$

The phase angle ϕ can have many values originating from various aberrations within the optical system. For our purposes, we shall continue to assume that the exposure system is nearly perfect so that the phase term is only dependent on the focus condition. In this case, ϕ can be approximated by the expression

$$\phi(z) \cong \pi z \lambda/p^2, \tag{10}$$

where z is the distance from the plane of best focus. A coherent imaging system characteristically has a magnitude of modulation of 1.0, but there is an additional phase dependency. The intensity of the coherent aerial image is given by $I(x) = |A(x)|^2$,

$$I(x) = \tfrac{1}{4}I_{max}[1 + (8/\pi) \cos \phi \sin(2\pi\nu x) + (4/\pi)^2 \sin^2(\pi\nu x)]. \tag{11}$$

Equations (10) and (11) are equally valid for projection printers with coherent illumination and contact/proximity printers using collimated illumination. In the latter case, parameter z is replaced by y, the separation distance between the mask and wafer.

Comparing Eq. (11) with Eq. (8), we note that, for a given I_{max}, the average light intensity with incoherent illumination is twice the average intensity that will occur in the coherent case. In the sections that follow, we shall observe that this factor of 2 plays a major role in explaining differences in profile angle and linewidth control.

C. Resist Profile

Having derived the shape of the aerial image for incoherent and coherent imaging, we are now ready to examine the shape of the resist profile that will result with a specific resist–process–substrate combination. Specifically, we want to know the position and the slope of the profile at the point where the resist thickness falls to zero. The characteristics of the resist profile in this region ultimately determine the quality of the image transfer for either wet or dry etching processes.

Consider the case of a resist profile similar to that shown in Fig. 1, where the resist is T_0 micrometers thick and where the resist, process, substrate, and exposing wavelength combination is characterized by a specific γ and threshold E_0. To calculate the slope of the resist profile, we follow the approach of Blais [11] by expressing the slope of the resist profile as the product of two partial derivatives:

$$dT/dx = \tan \theta = (\partial T/\partial E)\,(\partial E/\partial x). \tag{12}$$

The term $\partial T/\partial E$ is mainly dependent on the process. The term $\partial E/\partial x$ is the slope of the aerial image and is a function of the exposure tool and the mask pattern.

Following the approach of Bruning and Shankoff [12], the process-dependent term can be calculated from the characteristic curve. From the definition of γ in Eq. (2), the resist thickness remaining in the vicinity of the threshold for $E < E_0$ is given by

$$T(E) = T_0\gamma \log_e(E_0/E), \tag{13}$$

where T is the unexposed resist thickness and $E(x)$ the exposure at position x. By differentiating with respect to E, the slope is given by

$$\partial T(E)/\partial E = -T_0\gamma/E. \tag{14}$$

The second term in Eq. (12) can be obtained from Eq. (8) for the case of incoherent imaging. By multiplying both sides of the equation by the exposure time, we can write Eq. (8) in terms of exposure

$$E(x) = \tfrac{1}{2}E_M[1 + (4/\pi)M_I(\nu)\,\sin(2\pi\nu x)]. \tag{15}$$

Here the term E_M corresponds to what we normally refer to as the "exposure." This is the exposure that would occur from a large clear area on the mask. From this expression we obtain

$$\partial E/\partial x = 4E_M\nu M_I(\nu)\,\cos(2\pi\nu x). \tag{16}$$

Substituting Eqs. (14) and (16) into Eq. (12),

$$\tan_I(\theta) = 4\gamma T_0[E_M/E(x)]\nu M_I(\nu) \cos(2\pi\nu x). \tag{17}$$

Recall that Eq. (15) is a function of a series representing a pattern of equal lines and spaces. The position $x = 0$ corresponds to the point of transition between a clear and an opaque region on the mask. When exposure E_M is adjusted such that $E(0) = E_0$, the resist pattern will delineate equal lines and spaces on the wafer. From Eq. (15) we see that this will occur when $E_M = 2E_0$. Under these conditions we obtain

$$\tan_I(\theta) = 8\gamma T_0\nu M_I(\nu). \tag{18}$$

Equations (17) and (18) have been found by the author to give excellent agreement with experiment. As discussed in Section V.D, the product γT_0 is independent of T_0 when T_0 is large (typically greater than 0.5 μm). However, γT_0 becomes proportional to T for very thin resist layers. For this reason, steep profile angles are easier to obtain when the resist is thick. A similar result has been reported by Tigreat [13]. Note that the feature size enters Eq. (18) only in the product $\nu M(\nu)$. By setting $E(x) = E_0$ in Eq. (17) we see that $\tan_I(\theta)$ to the first order is proportional to the exposure E_M.

Starting with Eq. (11), similar expressions for the resist profile angle can be derived for the case of coherent illumination:

$$\tan_C(\theta) = 4\gamma T_0[E_M/E(x)]\nu \cos(2\pi\nu x)[\cos \phi + (4/\pi) \sin(2\pi\nu x)] \tag{19}$$

and

$$\tan_C(\theta) = 16\gamma T_0\nu \cos \phi \tag{20}$$

when the exposure is adjusted to obtain equal lines and spaces on the wafer, which occurs when $E_M = 4E_0$. Comparing our results for coherent and incoherent illumination, we conclude that the slope of the resist profile at the resist–substrate interface is more than a factor of 2 greater and proportional to the spatial frequency (assuming perfect focus) in the case of coherent illumination.

Calculating the resist profile angle for the case of partial coherence is beyond the scope of this chapter. However, it is useful to take a heuristic approach by writing

$$\tan_P(\theta) = 8Q\gamma T_0\nu M_P(\nu) \tag{21}$$

in place of Eq. (18). Here M_P is the modulation obtained from the MTF curves in Fig. 11 and $Q = E_M/2E_0$, when E_M is the exposure found to achieve equal lines and spaces on the wafer. The value of Q is found to increase monotonically from 1.0 to 2.0 with increasing coherence.

V. FEATURE SIZE CONTROL

To preserve device yield, it is essential to maintain the correct width of the feature, not only over the entire wafer, but for all wafers in a lot, and from lot to lot. Indeed, the ability to control linewidth is usually the factor that determines the limiting feature size. It depends on variations in exposure energy, changes in the resist threshold, and departures from perfect focus. Numerous authors have calculated linewidth variations that occur under specific conditions of exposure, wafer composition, and process. With the use of computer modeling, it is possible to predict accurately the changes in resist profile that occur as a result of specified changes in exposure, substrate, development, etc. [14]. In this section, we shall use our simplified model to derive the functional relationships that exist between feature size control and the lithographic variables.

A. Linewidth Control with Incoherent Illumination

In order to determine the linewidth dependence on each lithographic parameter, for the case of incoherent illumination, we can solve Eq. (15) for x and calculate

$$\frac{\Delta L}{L} = \frac{2}{L}\frac{\partial x}{\partial E_M}\Delta E_M + \frac{2}{L}\frac{\partial x}{\partial E}\Delta E + \frac{2}{L}\frac{\partial x}{\partial M}\Delta M. \tag{22}$$

We again limit our analysis to the simplified case of imaging an equally spaced grid of spatial frequency ν, where only the fundamental frequency plays a significant role. Considering the fractional change in linewidth for a resist pattern corresponding to an opaque feature on the mask, we evaluate Eq. (22) at the position $x = 0$:

$$\left(\frac{\Delta L}{L}\right)_I = -\frac{1}{M_I}\cdot\frac{E}{E_M}\frac{\Delta E_M}{E_M} + \frac{1}{2M_I}\frac{\Delta E}{E} - \frac{1}{2M_I}\left(\frac{2E}{E_M} - 1\right)\frac{\Delta M_I}{M_I}. \tag{23}$$

When the exposure is adjusted to obtain equal lines and spaces on the wafer, $E_M = 2E_0$ and $E(0) = E_0$. Substituting these values into Eq. (23), we obtain

$$\left(\frac{\Delta L}{L}\right)_I = -\frac{1}{2M_I}\left(\frac{\Delta E_M}{E_M} - \frac{\Delta E_0}{E_0}\right). \tag{24}$$

The first term in Eq. (24) describes the functional relationship involving modulation, nonuniformity of the illumination, and variations in the linewidth. For example, if our exposure tool has a nonuniformity of illumination of $\pm 10\%$ and we require a $\pm 10\%$ linewidth control, we are restricted to features for which the modulation is greater than 50%.

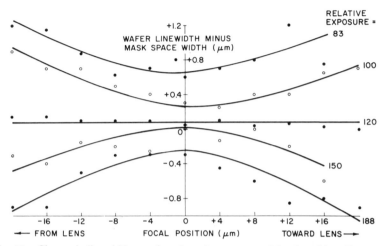

Fig. 12. Change in linewidth as a function of exposure and focal position. Feature size was a 2.0-μm line and space on the mask. The exposure system was an $f/3$ 1:1 projection printer, $\lambda = 3900$ Å, $S = 0.857$, exposing in 1.0 μm of AZ-1370 resist. Note that the feature size is insensitive to focus for the exposure that achieves the nominal feature size.

The second term in Eq. (24) describes the functional relationship involving modulation, variations in resist threshold exposure, and variations in linewidth. As previously defined, E_0 is a function of all the process- and substrate-related parameters. As in the case of exposure nonuniformity, the greater the modulation, the greater the latitude for process variations.

The third term in Eq. (23) does not appear in Eq. (24) because it vanishes when $E(0) = E_M/2$. This expression describes the variations in linewidth that occur when the optical system departs from optimum focus. The functional dependence of $\Delta M/M$ on focus can be calculated from Eq. (5) and Fig. 10. More importantly, we see that by choosing the exposure to be twice the resist threshold (which is the same condition for equal lines and spaces[1]) the ability to control linewidth becomes independent of focal position. This can be verified experimentally, as shown in Fig. 12, where ΔL is measured as a function of focal position and exposure. Here, as predicted by Eq. (23), the linewidth becomes insensitive to focus for exposures that reproduce the nominal linewidth. Also note that the slopes of the curves for exposures greater than and less than $2E_0$ are correctly

---

[1] In the general case when $L \neq p/2$, the exposure for nominal feature size will differ from the operating point exposure by the multiplicative factor $[1 + (4/\pi)M_1 \cos(\pi L/p)]$ for incoherent illumination and by $[1 + (8/\pi) \cos(\phi) \sin(\pi L/p)]$ for coherent illumination.

predicted. The existence of operating points at which linewidth can be maintained independent of focus was first described by Cuthbert [5] by calculating the image of a step function under various conditions of focus and partial coherence.

By combining the rms values from Eq. (24), we obtain

$$\left(\frac{\Delta L}{L}\right)_{\mathrm{I}} = \frac{1}{2M_{\mathrm{I}}} \left[\left(\frac{\Delta E_{\mathrm{M}}}{E_{\mathrm{M}}}\right)^2 + \left(\frac{\Delta E_0}{E_0}\right)^2 \right]^{1/2} \tag{25}$$

for the fractional linewidth variation caused by variations in exposure and processing. This expression demonstrates clearly that large image modulation is the most important factor in achieving good linewidth control. Increasing the modulation can reduce the effect of exposure and processing variations. Likewise, when smaller features are imaged, linewidth control becomes more difficult by virtue of the lowered modulation. Equation (25) points out the various tradeoffs that can be used to obtain the necessary linewidth control. For example, to achieve $\pm 10\%$ linewidth control at 60% modulation with a nonuniformity of exposure of $\pm 10\%$ leaves a $\pm 6\%$ allowable variation for the processing. Similarly, if we increase the modulation to 80% and assume a 5% nonuniformity of exposure, the tolerance for process variations is increased to 15%.

Equation (25) contains a subtle relationship between focus and linewidth control. By exposing at the operating point and dropping the third term in Eq. (23), the expression is still dependent on focus as a result of M in the denominator. When the system becomes defocused, the modulation decreases, which leads to less latitude for exposure and process variations. As will be established in Section VI, a practical focal tolerance is a depth of focus that maintains the modulation above 40%.

B. Linewidth Control with Coherent Illumination

Starting with Eq. (11) and substituting the various partial derivatives into Eq. (22), we can calculate the fractional variation in linewidth that will occur when using coherent illumination. With the same mask pattern used in the previous discussion, we obtain

$$\left(\frac{\Delta L}{L}\right)_{\mathrm{c}} = -\frac{1}{\cos\phi} \cdot \frac{E}{E_{\mathrm{M}}} \frac{\Delta E_{\mathrm{M}}}{E_{\mathrm{M}}} + \frac{1}{4\cos\phi} \frac{\Delta E}{E}$$
$$-\frac{1}{4\cos\phi} \left(\frac{4E}{E_{\mathrm{M}}} - 1\right) \frac{\Delta\cos\phi}{\cos\phi}. \tag{26}$$

From Eq. (11) we note that equal lines and spaces will occur on the wafer when exposure E_M is adjusted such that $E(0) = E_0$, which implies that $E_M = 4E_0$. Substituting these values into Eq. (26), we obtain

$$\left(\frac{\Delta L}{L}\right)_C = -\frac{1}{4 \cos \phi} \left[\frac{\Delta E_M}{E_M} - \frac{\Delta E_0}{E_0}\right]. \tag{27}$$

Comparing the fractional change in linewidth using coherent illumination to the situation that occurs with incoherent illumination, we observe three major differences. First, with coherent illumination, the operating point is at $E_M = 4E_0$, which is twice the exposure required for incoherent illumination. Fortunately, this is also the exposure value required to maintain the nominal linewidth. Second, the modulation has been replaced by cos ϕ, which has the value of 1.0 for perfect focus but decreases with focal error. Third, there is an additional value of 2 in the denominator, which implies that for the same exposure and process tolerances a coherent imaging system will exhibit half the percentage change in linewidth that would occur with incoherent imaging. This last difference is most significant. It points out the key advantage of contact printing and explains why the transition from contact printing to projection printing has often required major improvements in processing before an increase in yield can be realized.

C. Linewidth Control with Partial Coherence

In the real world, we are interested in projection systems that have a certain degree of partial coherence. Calculating linewidth control as a function of partial coherence is beyond the scope of this chapter. Again, we will take an heuristic approach by writing

$$\left(\frac{\Delta L}{L}\right)_P = \frac{1}{2QM_P} \left[\left(\frac{\Delta E_M}{E_M}\right)^2 + \left(\frac{\Delta E_0}{E_0}\right)^2\right]^{1/2}, \tag{28}$$

where Q is the ratio of the measured operating point exposure with partial coherence to the operating point exposure with incoherent illumination. The operating point occurs at the exposure $E_M = 2QE_0$. The relationship between linewidth control and partial coherence has been treated in detail in a paper by Lacombat and Dubroeucq [15]. Their results agree quite well with those predicted by Eq. (28).

D. Linewidth Control over Steps

An important aspect of the linewidth control problem is the situation that occurs when imaging over steps. In general, the resist thickness will undergo large fluctuations whenever there are sharp changes in substrate

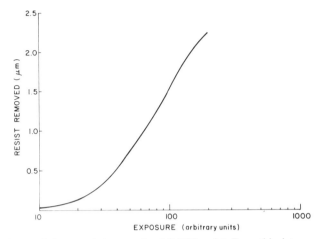

Fig. 13. Inverted characteristic curve for AZ-1300 resist. From this data, γ can be calculated for a given resist thickness.

topography. This makes it difficult to control the width of the resist profile, for example, when it passes from a thick to a thin oxide layer. The origin of this effect can be understood from Fig. 13, which is an "inverted characteristic curve" for the AZ-1300 resist. From this curve, γ can be obtained for any specified resist thickness. The slope of the curve is constant over the range 0.5 μm $\leq T_0 \leq$ 2.0 μm. If the resist thickness is maintained in this range, γT_0 is a constant, which implies that

$$\Delta\gamma/\gamma = -\Delta T_0/T_0. \tag{29}$$

From Eq. (2),

$$\Delta E_0/E_0 = -(1/\gamma)(\Delta\gamma/\gamma). \tag{30}$$

Substituting Eqs. (29) and (30) into Eq. (28), we obtain

$$(\Delta L/L)_P = (1/2M_PQ\gamma T_0)\,\Delta T_0 = 2\,\Delta T_0/L\,\tan_P(\theta). \tag{31}$$

From this expression we see the importance of having steep profile angles when imaging over steps. This emphasizes the need for good modulation, large γ, and the use of partial coherence when variations in resist thickness are expected. The same conclusion was found by Widmann and Binder [16] using a different approach to the problem.

This treatment has ignored the effect of standing waves. When the substrate reflects a large percentage of the illumination and the change in resist thickness allows more or less light to become coupled into the resist film, substantial changes in E_0 can occur. Cuthbert [5] has calculated this effect for a number of cases.

E. Linewidth Control with Dry Etching

In the previous section we developed the functional relationship between the key lithographic parameters and the control of the resist profile width at the resist–substrate interface. For device fabrication, we are ultimately interested in the uniformity of the etched pattern. When the resist is impervious to the etch, as in the case of wet etching, the relationships presented in Eqs. (25), (27), and (28) are adequate.

Consider the situation depicted in Fig. 14, where an anistropic etch removes a resist thickness h during the process of image transfer. As we see in the figure, the edge of the resist profile will move by an amount

$$u = h/\tan \theta \tag{32}$$

and the width of the resist profile will be narrower by the amount $2u$. In order to etch the nominal linewidth into the substrate, we shall plan to underexpose by an amount necessary to compensate for the loss in feature size. This is accomplished by requiring that the resist thickness T remaining at $x = 0$ (location of the nominal feature edge) be equal to h, the resist removed by the etch. From Eq. (13),

$$h = \gamma T_0 \log_e[E_0/E(x)]. \tag{33}$$

By solving this expression for $E(x)$, substituting the value into Eq. (33) for the incoherent aerial image, and solving for x to the first order, we obtain

$$x = (L/4M_I)[(2E_0/E_M)e^{-h/\gamma T_0} - 1]. \tag{34}$$

The fractional change in linewidth is given by

$$\begin{aligned}
\frac{\Delta L}{L} = \frac{2}{L} \Bigg(&\frac{\partial x}{\partial M_I} \Delta M_I + \frac{\partial x}{\partial E_0} \Delta E_0 \\
&+ \frac{\partial x}{\partial E_M} \Delta E_M + \frac{\partial x}{\partial h} \Delta h + \frac{\partial x}{\partial \gamma T_0} \Delta \gamma T_0 \Bigg),
\end{aligned} \tag{35}$$

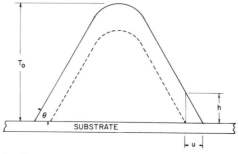

Fig. 14. Schematic of a resist profile (———) before and (– – –) after an anisotropic etch. Removing the resist by the amount h results in an edge loss u.

from which we obtain

$$\left(\frac{\Delta L}{L}\right)_{\mathrm{I}} = \frac{2}{L}\left[\frac{-L}{4M_{\mathrm{I}}^2}\left(\frac{2E_0}{E_{\mathrm{M}}}\,e^{-h/\gamma T_0} - 1\right)\Delta M_{\mathrm{I}} + \frac{Le^{-h/\gamma T_0}}{2M_{\mathrm{I}}E_{\mathrm{M}}}\,\Delta E_0\right.$$

$$\left. - \frac{Le^{-h/\gamma T_0}}{2M_{\mathrm{I}}E_{\mathrm{M}}^2}\,\Delta E_{\mathrm{M}} - \frac{LE_0e^{-h/\gamma T_0}}{2M_{\mathrm{I}}\gamma T_0 E_{\mathrm{M}}}\,\Delta h\right]. \tag{36}$$

We have assumed that $\Delta\gamma T_0 = 0$. Evaluating Eqs. (34) and (36) at $x = 0$, we obtain

$$E_{\mathrm{M}} = 2E_0 e^{-h/\gamma T_0} \tag{37}$$

for the correct exposure, and Eq. (36) reduces to

$$\left(\frac{\Delta L}{L}\right)_{\mathrm{I}} = \frac{1}{2M_{\mathrm{I}}}\left(\frac{\Delta E_0}{E_0} - \frac{\Delta E_{\mathrm{M}}}{E_{\mathrm{M}}}\right) - \frac{2\,\Delta h}{L\tan\theta}, \tag{38}$$

where

$$\tan\theta = 4T_0\gamma\nu M_{\mathrm{I}}(E_{\mathrm{M}}/E_0 e^{-h/\gamma T_0}) = 8T_0\gamma\nu M_{\mathrm{I}} \tag{39}$$

is the slope of the resist profile.

Comparing Eq. (38) with Eq. (24), we see that under the stated conditions the fractional change in linewidth is equal to the change associated with the lithography alone plus an additional etch-dependent term that is dependent on Δh but independent of h. This expression stresses the need for steep profile angles when etching small features.

From this analysis we conclude that underexposing to maintain the nominal feature size is of critical importance for achieving good linewidth control. Under this exposure condition, the first term in Eq. (36) vanishes, allowing us to maintain the full focal tolerance (see Section VI).

VI. FOCAL TOLERANCE

The allowable depth of focus for a lithographic exposure system depends on our criteria for acceptable performance. The incoherent illumination, the deciding factor, is determined by the ability to maintain a desired degree of linewidth control. In the presence of nonuniformities in process and exposure, we can relate the depth of focus to the minimum value of the image modulation, which will allow us to maintain a 10% linewidth control. Assuming that exposure is at the operating point, from Eq. (25) we can write

$$M_{\mathrm{I}} \geq (1/0.2)[(\Delta E_{\mathrm{M}}/E_{\mathrm{M}})^2 + (\Delta E_0/E_0)^2]^{1/2} \tag{40}$$

for incoherent illumination.

Considering today's level of process control, 40% is usually accepted as the minimum allowable modulation M_l that enables one to maintain 10% linewidth control. From Fig. 11 and Eq. (5) we can calculate the depth of focus for an incoherent system. By fitting a linear curve to D^2 in Fig. 11, assuming the unaberrated MTF to be greater than 60% for a periodic feature of period $2L$, and allowing the modulation to fall to 40%, one can show (Appendix) that the depth of focus δ_1 is given by

$$\delta_1 = \pm 1.85 fL. \tag{41}$$

We conclude that for incoherent illumination the depth of focus is proportional to the f/number f and the feature size L but independent of wavelength. A physical interpretation of the independence of the incoherent depth of focus on wavelength has been previously discussed by King [17].

In the case of coherent illumination, linewidth control is easier to maintain. However, as pointed out by Dubroeucq and Lacombat [18], the depth of focus is limited by the appearance of unwanted diffraction effects that are a function of the feature size and the feature configuration. In the special case of equal lines and spaces, the effect of defocus on the aerial image can be studied by substituting different values for ϕ in Eq. (11). As the distance from the plane of best focus increases, the magnitude of the intensity maxima decreases. In addition, secondary maxima begin to appear in the regions that represented a minimum intensity in the focused aerial image. The resultant effect is an aerial image with twice the spatial frequency of the mask pattern. For a thorough discussion of this effect the reader is referred to a paper by Tigreat [13].

An expression for acceptable depth of focus performance with coherent illumination is obtained by requiring that the exposure value at the secondary maxima does not exceed the threshold exposure of the resist. Assuming a nominal exposure to achieve equal lines and spaces in the developed resist, we can express this condition by the expression

$$E(x = -1/4\nu) \leq E_0 = E(x = 0). \tag{42}$$

By substituting Eq. (11) into Eq. (42), we obtain

$$\cos(\phi) \geq 0.64; \tag{43}$$

substituting this value in Eq. (10) we obtain the depth of focus for a coherent optical system:

$$\delta_C = \pm 1.1 L^2/\lambda. \tag{44}$$

Here we find that the depth of focus is a function of the wavelength and feature size but independent of f. For contact/proximity printers, δ_C is the maximum proximity gap z that will allow one to obtain minimum features of dimension L. In practice, the allowable proximity gap would be less as a result of a decollimation of the illumination. By equating Eq. (41) to Eq. (44) we see that for features greater than $L = 1.68$, λf coherence provides an enhanced depth of focus.

A similar calculation for partial coherence is beyond the scope of this chapter. From Eqs. (41)–(44) the effects of partial coherence can be inferred. By increasing the partial coherence, the depth of focus becomes more dependent on λ in a favorable way, less dependent on f, and more dependent on the feature size in an unfavorable way.

In Fig. 15, the depth of focus is plotted for coherent and incoherent illu-

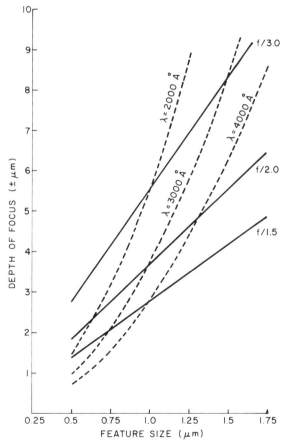

Fig. 15. Calculated depth of focus for small features as a function of coherence. The depth of focus is independent of λ for incoherent systems (—) and independent of f/number for coherent systems (– – –). The only values that should be considered in this figure are for feature size, f/number, and wavelength combinations that provide a modulation greater than 60% in the focused state as determined from Fig. 11 (see guideline in text).

mination as a function of feature size for three values of λ for coherent illumination, and three values of f/number for incoherent illumination. The spread between the coherent depth of focus for a given λ and the incoherent depth of focus for a given f is an indication of the benefits that can be achieved with partial coherence. The reader is cautioned that Fig. 15 contains f and feature size combinations for which the unaberrated MTF falls below 60% and are therefore invalid for consideration.

Our calculation of depth of focus is based on the assumption that the minimum allowable modulation for incoherent imaging is 40%. Assuming that process control will improve by the use of multilevel resist technology, it is useful to examine how far the modulation can be allowed to fall. A reasonable limit is imposed by requiring that the resist thickness not be diminished as a result of low profile angles. This requires that tan $\theta \geq 2T/L$. From Eq. (17) we obtain

$$M_I \geq 1/2\gamma \tag{45}$$

for incoherent illumination and

$$M_P \geq 1/2Q\gamma \tag{46}$$

with partial coherence. We therefore conclude that greater depth of focus can be achieved by using resist systems with large γ. Unfortunately, the use of multilayer resist technologies with thin layers of conventional resist involves low γ systems. The use of multilayer resist technologies with inorganic photosensitive layers may offer some relief.

VII. OVERLAY

Thus far we have limited our discussion to the quality of lithographic imagery. Of equal importance is the accuracy with which the image can be positioned over the surface of the wafer. During the process of fabricating devices, each level must be aligned to the previous levels. Most IC design rules call for an overlay tolerance no greater than a small fraction of the minimum feature size. From the standpoint of exposure tool design, overlay, not linewidth control, commonly becomes the limiting characteristic of the exposure technology.

The total overlay error budget is composed of many terms. The number of terms increases as the overlay tolerance becomes tighter. Today, state-of-the-art lithography is operating with minimum working features between 1.0 and 2.5 μm. In this region device designers require an overlay tolerance in the 0.25- to 0.75-μm range (2σ). To achieve this level of performance, very tight control over many parameters is required.

A. The Exposure Tool

The characteristics of the exposure tool that affect overlay are: the design of the optical system that allows viewing of the relative positions of the image and wafer before exposure, the means for manipulating the position of the image on the wafer, the thermal stability of the mask and wafer substrates, and the geometrical distortion in the optical image. In addition, we must consider the distortions introduced by the use of different masks, by the process itself, and by chucking the wafer against a flat chuck.

In a contact/proximity printer, the mask and wafer are viewed through the back of the mask with a high-powered microscope. To view the mask and wafer simultaneously requires bringing the two surfaces in close contact. To move the wafer requires increasing the separation to at least 10 μm. The alignment procedure is an iterative process requiring a number of contact and separation operations prior to exposure. This exaggerates the problem of mask and wafer damage inherent in a contact printing operation.

With projection printing, the alignment system is more complex. As in the case of contact printing, the illumination must be outside the actinic bandwidth ($\lambda > 4500$ Å). This places an added burden on the optical design of refractive systems and complicates the coating designs for all-reflective systems. The tendency today is to move away from manual alignment toward automatic alignment systems employing pattern recognition or signal processing. Automatic alignment offers a number of advantages other than speeding up the alignment process and achieving better accuracy. By performing real-time calculations, an automatic alignment system can estimate linear magnification errors. This can be utilized on exposure equipment that allows for real-time correction of magnification errors. It can also make corrections for systematic alignment errors produced by mechanical offsets and optical distortions within the viewing system. With automatic alignment, 0.1-μm alignment accuracy will be obtainable in the future. Alignment should not impose a serious limitation on submicrometer VLSI performance goals.

Full-field exposure systems align to two positions on the wafer. On the other hand, steppers have the inherent capability to align every die if necessary. Die-by-die alignment is an attractive feature but tradeoffs with throughput will require careful consideration.

B. The Effect of Vibration on Alignment

Once good alignment is achieved, it cannot be assumed that it will be maintained during the exposure process. Low-frequency vibration (≤ 10

Hz) can easily vary the relative positions of the mask/reticle and the wafer. These vibrations have little effect on imagery but increase the statistical spread of the correctly aligned die. Low-frequency vibrations are difficult to filter out. When pursuing submicrometer VLSI, increased efforts will be required to limit environmental vibration.

C. Optical Distortion

It is common for many optical systems to distort an image in a manner that is symmetrical about the optical axis. This difficulty becomes more severe as the image field size is increased. The inherent symmetry in a 1:1 projection system allows the design of optics that are free of distortion. Refractive reduction systems always have some residual distortion. State-of-the-art lens designs have been able to hold distortions under 0.1 μm.

Often, the greatest source of optical distortion is imperfections in the optical elements and assembly. The performance of an optical system is measured by the optical path differences (OPD) encountered by two waves passing through the system. An OPD states by how much a wavefront measured in wavelengths deviates from an idealized case. Lithographic projection systems require diffraction-limited performance, which mandates that any ray passing through the system can have an OPD no larger than $\lambda/4$. Localized optical distortions are not affected by the magnitude of the OPD errors but rather by the slopes on the OPD surface. The OPD surface is a plot of the OPD values as a function of position in the image plane. A small tilt error on an optical surface that introduces an OPD = $\lambda/4$ can still displace part of the image by more than a micrometer. A guarantee that localized optical distortions are less than 0.25 μm thus often requires an order of magnitude decrease in fabrication tolerances, putting a considerable burden on the design and fabrication of lithographic exposure systems. Fortunately, advances in optical fabrication technology have kept pace with the requirements imposed by VLSI design rules. This trend is expected to continue. Levels of distortion consistent with the requirements of submicrometer design rules should be achievable in the foreseeable future.

D. Wafer Distortion

When examining the extension of full-field wafer exposure technologies, there is a major concern regarding wafer distortion. During the various processes that take place between exposure levels, the wafer can undergo an expansion or contraction that can easily exceed 1.0 μm. This can have a critical effect on the ability to achieve tight registration accuracy over the entire wafer. If the effect is linear and repeatable from wafer to wafer, it can be compensated for at the mask-making level or in the exposure tool. However, if a significant error is nonrepeatable or nonlinear, good overlay would require die-by-die alignment.

Cuthbert [19] has made an extensive study of in-plane distortions that occur during a state-of-the-art polysilicon NMOS process. He found the rms deviations from linear distortion to be less than 0.1 μm. The repeatability of the linear term was better than the limits imposed by his measurement technology [$\pm$3.7 ppm (1σ) over a distance of 2 in.]. Other investigators have reported similar results, but there is still considerable work to be done in this area. The experiments are difficult to perform and the experimental errors often mask the desired results. It now appears that by using a high degree of process control it will be possible to achieve 1.0-μm design rules using full-field exposure. For submicrometer lithography, some form of die-by-die alignment will probably be required.

E. Other Overlay Contributions and Considerations

There are other sources of overlay error to be considered in the total error budget. Mask-to-mask runout introduced at the mask-making step can easily introduce 0.25-μm errors. Considering the coefficient of thermal expansion for silicon (2.5 $\times$ 10^{-6}/°C), a temperature variation of 0.5 °C can introduce a runout of 0.13 μm over a 4-in. wafer. Both steppers and 1:1 projection printers can distort the imagery by misadjustment of their mechanical configurations.

Another important consideration involves the intermixing of different exposure tools. If all device levels are exposed on the same exposure tool, it is possible to eliminate the contribution of repeatable machine distortion. Today it is common practice to intermix many exposure tools of the same type. In the future it is possible that different types of optical exposure equipment will be intermixed among themselves and even with nonoptical technologies. This practice will place an additional burden on the overlay tolerance.

VIII. CONCLUSION

Summarizing the results of the previous sections on linewidth control, focal tolerance, and overlay, it becomes clear that the conditions needed for optimum lithographic performance are: large image modulation, a high degree of partial coherence, large resist γ, and the largest acceptable f/number in combination with the shortest possible wavelength. In addition, submicrometer lithography will probably require a small-field optical system to maintain overlay.

The extension of optical lithography to the submicrometer realm will require two major developments. From the standpoint of the resist, it will require higher-γ systems that can be used at shorter wavelengths. From the standpoint of the exposure equipment, it will require small-field systems capable of exposing in the deep UV with a high degree of partial coherence. The need for partial coherence will call for very tight manufacturing tolerances to reduce the phase-related aberrations that occur with coherent light and more intense illumination sources to maintain throughput.

APPENDIX

Effect of f/Number and Wavelength on the Incoherent Focal Tolerance[2]

In a system in which the unaberrated MTF is greater than 0.6,

$$M_0 = 1 - 1.22\nu\lambda f,$$

where M_0 is the unaberrated MTF, ν the spatial frequency, λ the wavelength, and f the f/number. In such systems, $\nu\lambda f \leq 0.33$.

For defocus $= z$, we have

$$R = z/2f^2\lambda, \qquad (A.1)$$

where R is the defocus in units of the Rayleigh tolerance. The decrease in MTF due to the defocus z is equal to $R^2 D^2(\nu\lambda f)$

For $\lambda f \leq 0.33$,

$$D^2(\nu\lambda f) = \nu\lambda f/3. \qquad (A.2)$$

The value of the MTF when the focus error is z is given by

$$M(z) = M_0 - R^2 D^2(\nu\lambda f) = 1 - 1.22\nu\lambda f - \frac{z^2}{4f^2\lambda^2}\left(\frac{\nu\lambda f}{3}\right) \qquad (A.3)$$

$$= 1 - 1.22\nu\lambda f - \frac{z^2}{12f^3\lambda}. \qquad (A.4)$$

---

[2] This Appendix was written by Abe Offner.

There is a minimum value of M_{min} of the MTF that is required for recording. For $M(\delta) = M_{min}$, we have

$$\delta^2 v/12 f^3 \lambda = 1 - M_{min} - 1.22 v \lambda f, \tag{A.5}$$

$$\delta^2 = 12(1 - M_{min}) f^2 \left[\frac{f\lambda}{v} - \frac{1.22}{1 - M_{min}} f^2 \lambda^2 \right] \tag{A.6}$$

$$= 12(1 - M_{min}) \frac{f^2}{v^2} \left[f\lambda v - \frac{1.22}{1 - M_{min}} f^2 \lambda^2 v^2 \right]. \tag{A.7}$$

For $M_{min} = 0.39$,

$$\delta = (2.7 f/v)[f\lambda v(1 - 2 f\lambda v)]^{1/2}, \tag{A.8}$$

where $f\lambda v < 0.33$. In practical systems, for imaging feature sizes less than 2 μm, if $2000 \le \lambda \le 4358$, then $0.15 \le f\lambda v \le 0.33$. The maximum value of $f\lambda v(1 - 2 f\lambda v)$ is reached at $f\lambda v = 0.25$. For other values of $f\lambda v$ in this range, the variation of the value of $[f\lambda v(1 - 2 f\lambda v)]^{1/2}$ is quite small as can be seen from the accompanying tabulation.

$f\lambda v$	$[f\lambda v(1 - 2f\lambda v)]^{1/2}$
0.33	0.335
0.25	0.354
0.20	0.346
0.15	0.324

In a practical, unaberrated system, the focus error that reduces the MTF to 0.39 for a feature size $L = 1/2v$ is given by the expression

$$\delta_1 = 1.85 fL. \tag{A.9}$$

The permissible focal error for a particular feature size is thus independent of the wavelength and directly proportional to the f/number for systems in which $L \le 2$ μm, 0.2 μm $\le \lambda \le 0.436$ μm, and $0.15 \le f\lambda v \le 0.33$.

ACKNOWLEDGMENTS

The author would like to acknowledge many useful discussions with C. P. Ausschnitt, J. L. Kreuzer, and A. Offner. The author would also like to thank A. M. Garber and A. W. McCullough for providing the experimental data.

REFERENCES

1. J. Kosar, "Light-Sensitive Systems," p. 104. Wiley, New York, 1965.
2. G. W. W. Stevens, "Microphotography," p. 2. Wiley, New York, 1968.
3. D. A. Markle, *Solid State Technol.* **17** (6), 50 (1974).
4. D. A. McGillis and D. L. Fehrs, *IEEE Trans. Electron Devices* **ED-22** (7), 471 (1975).
5. J. D. Cuthbert, *Solid State Technol.* **20** (8), 59 (1977).
6. F. H. Dill, A. R. Neureuther, J. A. Tuttle, and E. J. Walker, *IEEE Trans. Electron Devices* **ED-22** (7), 456 (1975).
7. J. M. Moran and D. Maydan, *J. Vac. Sci. Technol.* **16** (6), 1620 (1979).
8. K. L. Tai, W. R. Sinclair, R. G. Vadimsky, J. M. Moran, and M. J. Rand, *J. Vac. Sci. Technol.* **16** (6), 1977 (1979).
9. Joseph W. Goodman, "Introduction to Fourier Optics." McGraw-Hill, New York, 1968.
10. A. Offner, *Photogr. Sci. Eng.* **23** (6), 374 (1979).
11. P. D. Blais, *Solid State Technol.* **20** (8), 76 (1977).
12. J. H. Bruning and T. A. Shankoff, Far-UV 1:1 projection printing, *Proc. Microcircuit Eng., Aachen* (1979).
13. P. Tigreat, *Proc. Soc. Photo-Opt. Instrum. Eng. Dev. Semicond. Microlithogr. IV* **174**, 37 (1979).
14. C. N. Ahlquist, W. G. Oldham, and P. Schoen, A study of a high performance projection stepper lens, *Proc. Microcircuit Eng., Aachen* (1979).
15. M. Lacombat and G. M. Dubroeucq, *Proc. Soc. Photo-Opt. Instrum. Eng. Dev. Semicond. Microlithogr. IV* **174**, 28 (1979).
16. D. W. Widmann and H. Binder, *IEEE Trans. Electron Devices* **ED-22** (7), 467 (1975).
17. M. C. King, *IEEE Transactions Electron Devices* **ED-26** (4), 711 (1979).
18. G. M. Dubroeucq and M. Lacombat, Private communication.
19. J. D. Cuthbert, Characterization of in-plane wafer distortions induced by double polysilicon NMOS processing, *Proc. Microcircuit Eng. Aachen* (1979).

Chapter **3**

Resolution Limitations
for Submicron Lithography

M. P. LEPSELTER
W. T. LYNCH

Bell Telephone Laboratories
Murray Hill, New Jersey

I. INTRODUCTION

Present (1980) lithography is primarily being used for LSI circuits with 2- to 4-μm design rules. In the future, narrower design rules can be expected, with micron and submicron technologies becoming important in 1982 and beyond. There are "existence proofs" that submicron linewidths are available today with a choice of several lithographies; however, the establishment of a competitive technology includes many trade-offs, including cost, yield (defect density), linewidth control, registration, and reliability. In this chapter we shall comment on some of the engineering or trade-off limitations but shall concentrate on the fundamental resolutions that can be achieved among the competitors for submicron lithography. Future trends will be predicted.

The obvious present contender for submicron lithography is UV exposure. Its dominant advantage is its low cost and high throughput, accented by its established learning curve in optical lithography. Resolution and depth of focus have been significantly improved by the adoption of scanning field projection or direct-step-on-wafer (DSW) exposure. Submicron patterning is possible, with a diffraction-limited resolution of ~ 0.3 μm. Although UV lithography will still dominate for the next few years, it will receive less attention in this chapter than the other contenders. Chapter 2 is devoted entirely to optical/UV lithography.

The new lithographies include proximity x-ray printing and direct writing by either electron beam or ion beam. All three of these technologies have achieved submicron patterns and none are primarily diffraction limited. The resolution capabilities of x rays are primarily limited by practical geometrical considerations, although the fundamental limit for all three technologies is due to primary or secondary electron effects within the resist and substrate layers. Masked systems with flood exposures by electrons or ions have their own practical problems.

All lithographies that use masks with a 1:1 mask/image ratio are also limited by the resolution and accuracy of the masks themselves. The fundamental limitations of an electron- or ion-beam column for writing masks are discussed in some detail in Section V.

An important subject in submicron technology, which is not considered in any detail in this chapter, is registration. As the design rules get narrower, all technologies must pay more and more attention to registration. Registration has a direct influence on circuit density and circuit reliability. The alignment accuracy should typically be about 25% of the feature width. Direct writing or DSW mask systems with registration at each step offer a natural solution to this requirement.

The references at the end of this chapter are not intended to be totally comprehensive or to give credit to "original" papers. Many of the papers listed are recent publications and review articles that have their own extensive list of references. Texts that discuss the general subject of radiation interaction with matter are also listed.

II. IDEAL PROPERTIES

To achieve submicron resolution with the tolerances required for production line VLSI, all aspects of the lithography must be under precise control. Some "ideal" properties for the resist, the exposure source, the lithography machine, and the exposed substrate are discussed next.

A. Resist

The resist should have high sensitivity, high contrast, high resolution, and high etch resistance.

A high-sensitivity resist will require only a low input dose and, therefore, a short exposure time. For conventional organic polymer resists, which use a solution development to create relief images, a high sensitivity is usually correlated with large molecular weights. In a tightly coiled state, the volume of the polymer molecule may be $(50 \text{ Å})^3$ to $(200 \text{ Å})^3$. The polymer molecule is entwined, however, with other molecules, and so its outer dimensions are extended. The resist developer tends to uncoil the molecules and the final geometry (and location) of the molecule after development affects the resolution. Smaller molecule sizes will give better resolution but also lower sensitivity. The statistical variation of resolution as a function of sensitivity is considered in Section VI.

A high-contrast resist should have a narrow molecular weight distribution (before exposure) and a sharp transition in a plot of final resist thickness (after development) versus exposure dose. The resist and its developer must be considered as a composite system in discussing contrast. The contrast (and the sensitivity) assume a particular source, a particular resist thickness, and a particular developing condition. High-contrast inorganic resists, which avoid many of the problems in developing organic polymer resists, are currently being studied.

High resolution implies high contrast but includes the consideration of pattern definition as well as uniform exposure. Changes in developer concentration of a few percent, in temperature by a few degrees, or in initial resist thickness by a few hundred angstroms can change the resolved pattern width by more than a tenth of a micron, even if the exposure is perfectly controlled. Swelling of the resist during development, with later contraction and distortion, will also affect linewidth resolution. A high-resolution resist will have a stiff rheological backbone for narrow lines, no flow during developing, and no stringers due to microelastic distortions during the development process. Smaller molecule sizes will improve resolution. Dry (plasma) developing, as opposed to wet chemical developing, perhaps in conjunction with inorganic resists, offers significant hope for reducing linewidth variations and distortions. Thinner films will always offer better resolution, but the defect densities will be higher. The resist filtering must exceed the film thickness.

Resist thickness differences of hundreds to thousands of angstroms are obtained at steps in the wafer substrate for typical spin-coated resists. Resulting patterns may show severe necking or flaring at the steps. Optical/UV exposure is particularly sensitive to resist and substrate layer variations because of interference effects. Ideally, a flat substrate and a uniform resist layer are desired.

The resist film must provide a high etch resistance for the subsequent pattern transfer to the underlying substrate. Any resist erosion at the edges during pattern transfer will affect the final feature width.

B. Exposure Source

The exposure source should have a high brightness, sufficient to expose (and align) a wafer in 2 min or less. The source intensity must be uniform in area and time. The output should have no uncorrectable chromatic (or energy) aberration if it is to go into a lens system. The energy absorption in the resist should be uniform with no scattering effects. The wavelength should be less than both the desired resolution and the resist thickness in order to avoid diffraction and interference effects.

The wavelength λ is given directly for UV exposure. For other lithographies, the wavelength in angstroms is $12.4/E$ for x-ray photon exposure, $0.39/E^{1/2}$ for e-beam exposure, and $(m/M)^{1/2}(0.39/E^{1/2})$ for ion-beam exposure, where E is the energy in kiloelectron volts, m the mass of an electron, and M the mass of the ion. Diffraction is not of importance for x-ray (0.5–6 keV), e-beam (5–30 keV), or ion-beam (10–300 keV) exposures. Diffraction dominates for UV exposure.

Primary and secondary scattering events (Section IV) dominate the fundamental resolutions for x rays, e beams and ion beams. Resolution for x rays is controlled by mask limitations and by the range of the secondary electrons created in the resist; for e-beam by scattering-induced beam spreading within the resist layer; and for ion-beam by scattering; however, with ion-mask systems, the scattering is magnified because of the divergence angle of the output from the mask, and by the gap between mask and wafer.

C. Lithography Machine

The lithography machine should offer a uniform spot or flood exposure, automatic and accurate alignment, a reasonable depth of focus ($\geq \pm 3\ \mu$m for a full-wafer exposure, $\geq \pm 0.5\ \mu$m for a limited field exposure with refocusing), a distortion-free image field (with accurate "stitching" for limited fields), high throughput (>30 wafers/hr), and low cost.

D. Substrate

The ideal substrate is flat, i.e., with no steps. The substrate should present a uniform impedance to the source, with no preferential absorption or reflection. This will optimize linewidth control. The substrate should act as a sink for the incident radiation, with no backscatter, reflection, or proximity effects. Finally, the wafer substrate should be insensitive to radiation damage from the source.

A single layer of resist over a partially processed silicon wafer will not satisfy the preceding requirements. Various bilevel, trilevel, and quadlevel schemes have been proposed. Figure 1 shows the Bell Laboratories' trilevel structure.

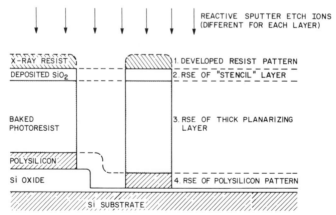

Fig. 1. Bell Telephone Laboratories' trilevel resist pattern. The sequence of steps from the initial exposure of the resist to the final pattern on the wafer is indicated.

→| |← 1.0 μm

Fig. 2. SEM micrograph of a polysilicon pattern produced by reactive sputter etching with the trilevel process of Fig. 1. The nearly vertical sidewall of the baked polymer layer and the insensitivity to steps in the silicon wafer are clearly indicated.

The bottom layer of baked resist acts as a surface-smoothing layer and as the etch-resistant layer for the final pattern transfer. It is not intended to be sensitive to the exposure from the source. The midlayer is a thin, stencil layer for transferring the pattern from the thin, high-resolution, top resist layer to the bottom layer. The top exposure-sensitive layer is spun onto a flat surface, so its thickness uniformity can be precisely controlled. Since there are no steps, the resist can be thin and the resolution can be improved. The resist must only be thick enough to provide an acceptable defect density.

The slice substrate is far enough ($\sim 2\ \mu m$) from the exposed resist that it does not affect the sensitivity. Exposure conditions do not have to be reset for each mask level, since the source always "sees" the same substrate. For e-beam or ion-beam exposure, the bottom layer can act as a sink for any residual radiation. Backscatter (Section V) to the sensitive resist layer is reduced by using a low atomic number organic material as the bottom layer.

After exposure and development of the top resist layer, the remaining structure is selectively and anisotropically etched by means of reactive sputter etching (RSE). The resist acts as a mask for the RSE of the stencil layer. The stencil layer, typically SiO_2 or polysilicon, has a high selectivity against etching while the organic bottom layer is etched. During the RSE of the bottom layer, the top resist layer, which is also organic, is etched away. Vertical sidewalls with no undercut are obtained in the bottom layer. This layer then acts as a thick (nearly defect-free) etching mask for the pattern transfer, usually by RSE, to the substrate. Figure 2 shows a completed structure.

The trilevel structure, in conjunction with RSE, offers as ideal a substrate structure as can be practically realized. Resist thickness is minimal, sidewalls are vertical, and resolution is maintained.

III. DIFFRACTION EFFECTS

A. Ultraviolet Effects

1. Contact Printing

Contact prints can be expected to give linewidths approaching $\lambda/2$ with an essentially faithful reproduction of the mask. Features of 0.25 μm have been imaged with 2000 to 2600 Å radiation. With contact printing, however, the defect density increases with each use of the mask. Registration is particularly difficult to control. Resolution may also vary across the silicon wafer depending on how well the mask and the wafer conform to each other.

2. Proximity Printing

Proximity printing avoids contact between the mask and wafer and, therefore, offers longer mask life. The resolution is a strong function of the gap spacing. In the range of Fresnel diffraction, the gap spacing s must be such that

$$\lambda^2 \ll s\lambda < W^2, \tag{1}$$

where W is the minimum slit width. The diffraction pattern is then a function of the system Q, where

$$Q = W(2/\lambda s)^{1/2}, \tag{2}$$

and the divergence angle α of the incident radiation. A large Q means that there are more Fresnel zones ($\propto W/\lambda$) within the slit width, with the probability for smoothing of the diffraction pattern beyond the width W in the image plane increasing as the number of zones or the angular spread of the slit ($\propto W/s$) increases. An increase in α will smooth the interference pattern both within and without the width W in the image plane. With $\alpha = 0$, i.e., coherence, the diffraction interference produces a lateral spatial oscillation (ringing) in the absorption by the resist. The profiles are smoothed by increasing the divergence angle to a few degrees, but this can give a shallower slope to the edge profile. Any loss in edge acuity is important in determining linewidth variations. For equally spaced lines and spaces, definition is also lost because of overlapping edge tails.

If a W of 0.5 μm is desired, and λ is reduced to 2000 Å, then the required gap spacing s must be ≤ 1.0 μm. This is an impractical spacing for a full-wafer exposure system since wafer distortions and mask bowing usually exceed 1.0 μm.

Although proximity printing is suitable for x-ray exposure ($\lambda = 2.5-50$ Å), it is not suitable for full-wafer UV exposure ($\lambda \gtrsim 2000$ Å) for linewidths less than 1.5–2.0 μm.

3. Projection Printing

Projection printing offers higher resolution than proximity printing and permits a large separation between the mask and the wafer. The lens imaging properties can be determined from the impulse response for a point source or the transfer function for a periodic sinusoidal or step function source. The periodic step function source may represent, for example, a grating, a sequence of lines and spaces of a specified period.

The diffraction pattern for a point source is the well-known Airy pattern, which has its first node at a radius of $1.2\lambda f \approx 0.6\lambda/\sin\theta$, where 2θ is the angle subtended by the lens exit pupil at the image point and f the f/number, or focal number, of the objective lens (f = focal length/pupil diameter). Two incoherent Airy patterns at a spacing of $1.2\lambda f$ can be just barely resolved, but the contrast ratio is only 0.74. A practical rule of thumb for the minimum line or space of an image (composed, for example, as a superposition of Airy patterns) is $1.5\lambda f$ for an incoherent source and λf for a partially coherent source. High-quality lens systems approach this limit.

The one-dimensional Fourier equivalent to the Airy diffraction pattern is the modulation transfer function for a spatially sinusoidal source grating. For a source function,

$$I(x) = I_0[1 + C\cos(2\pi\nu x)]; \qquad (3)$$

the lens transfer function

$$A(\nu) = |A(\nu)| \exp[i\phi(\nu)] \qquad (4)$$

is contained in the image response

$$I'(x) = I_0'\{1 + |A(\nu)|C \cos[2\pi\nu x + \phi(\nu)]\}. \qquad (5)$$

For a perfect lens, $|A(\nu)|$ varies nearly linearly from a value of 1 at $\nu = 0$ to a value of 0 at the cutoff period frequency $\nu_{co} = 1/\lambda f$. The modulation (contrast) transfer function is

$$\text{MTF} = |A(\nu)|(I_0'/I_0). \qquad (6)$$

Figure 3 shows the modulation transfer curve for sinusoidal gratings with incoherent and partially coherent illumination: R is defined as the ratio of the f/number for the projection optics to the f/number for the condenser optics, with $R = 1$ corresponding to full aperture exposure and incoherence and $R = 0.33$ corresponding, for example, to a one-third aperture exposure and partial coherence. Although the cutoff frequency is higher for incoherence, it offers no advantage since the resist development is usually set for a modulation of 0.6 or higher. The use of partial coherence gives better image contrast and edge definition for all lower spatial frequencies. Full (laser) coherence produces intense diffraction patterns with strong interference effects with neighboring patterns; however, a laser source with mechanical scanning to give a partially coherent illumination can provide an ideal source of high intensity and high efficiency. With $R = 0.33$, the resolved linewidth or space is $\sim \lambda f$.

Although a small f/number (i.e., a large numerical aperture or collection angle) is desired in order to have sharper resolution, a lens with a small f/number has a higher rate of degradation of the optical path differences as the imaging surface moves out of the focal plane. A smaller f/number therefore has a smaller depth of focus. The depth of focus varies as λf^2. For linewidths greater than the diffraction limit, the depth of focus is larger. Partial coherence increases the depth of focus.

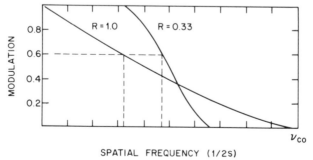

SPATIAL FREQUENCY (1/2S)

Fig. 3. Modulation transfer curve for periodic sinusoidal gratings of period $2s$ for incoherent ($R = 1.0$) and partially coherent ($R = 0.33$) illumination. The spatial frequency axis is normalized to the cutoff frequency $\nu_{co} = 1/\lambda f$.

B. X-Ray Effects

Diffraction effects are not significant for x rays (Section IV) except for wavelengths of 50Å or larger. Since there are no practical "optical" components for collimating or focusing x rays to the desired resolution (since the index of refraction $n \approx 1$), proximity printing is the only solution.

The relations given earlier for Fresnel diffraction also apply for x rays. In the worst case, $\lambda = 50$ Å, the spacing s from mask to wafer must be less than $W^2/50 \times 10^{-4}$. For $W = 0.2$ μm, for example, s must be less than 8 μm.

As long as direct proximity exposure is used, the spacings are within the range required by Fresnel diffraction, and there are no depth of focus effects.

C. Electron-Beam Effects

Diffraction effects are not significant for e beams (or ion beams). The de Broglie wavelengths are on the order of 0.1 Å or less. It is only the large f/number of 10^2 to 10^3 that keeps the diffraction resolution limit as high as 10–50 Å.

The depth of focus for electron- or ion-beam exposure is determined by geometrical considerations, specifically by the convergence angle α of the objective lens. The depth of focus is $\sim \pm 0.23 d_0/\alpha$, where d_0 is the minimum spot size, α is in radians, and defocusing is defined as $\pm 10\%$ in the spot size [see Eq. (14)]. For $d_0 = 0.2$ μm and $\alpha = 5 \times 10^{-3}$ rad, the depth of focus is $\sim \pm 9$ μm.

IV. SCATTERING AND ENERGY LOSS

A. Background

The ultimate resolution in any particular lithography is determined by the interactions of the incident particle/photon with the atoms and molecules of the resist. The resolution is determined by the scattering within the resist, since this determines the trajectory of the particle. The sensitivity is determined, in part, by the rate of energy loss to the resist. High sensitivity and high resolution (contrast) are usually incompatible. PMMA [poly(methyl methacrylate)] has the highest known resolution among common resists for e-beam lithography but also a low sensitivity.

Scattering may be elastic or inelastic. Although there is no energy loss in an elastic scattering event, later losses of energy along the altered trajectory directly affect the resolution. In an inelastic scattering event, the particle both loses energy and changes direction. The only energy loss events of interest are those that ultimately cause ionization, since it is the ionization that allows the restructuring of the resist. Polymer chain scission occurs for organic, positive resists, and polymer chain cross-linking occurs for organic, negative resists.

A secondary electron released by the scattering may itself have sufficient energy to cause further ionizations. Since its trajectory may be considerably different from the incident particle, the secondary electron can play a dominant role in determining the resolution. The secondary electrons also greatly affect the sensitivity of the resist since their rate of energy loss (energy loss per unit distance) is much greater than for the primary electron/photon, which has a higher energy.

Another factor in determining the resolution is the role of the substrate. Reflection from the substrate can produce interference patterns if the resist thickness is of the order of a few wavelengths or less. This is particularly important for optical/UV exposure. The incident particle may also enter the substrate and be backscattered into the unexposed areas of the resist. Particle backscattering is enhanced for high-density, high atomic number substrates where larger-angle scattering is more pronounced. The particles, as they enter the substrate, will have energies lower than their initial energies and this also enhances large-angle backscattering. Backscattering is of particular importance in e-beam exposure. Secondary electron generation is also enhanced in a high-density substrate; the short-range secondary electrons may enter the resist and produce a greater resist exposure within the first few hundred angstroms from the interface. Secondary electron effects at the interface are more pronounced for x-ray exposure, although the resolution is not necessarily affected. The ideal situation would be a nonreflecting, or vanishingly thin, substrate. Patterns with dimensions of 100 Å have been obtained for lines on thin silicon substrates. The trilevel scheme (Fig. 1) provides a practical compromise to the ideal substrate.

The factors affecting resolution, therefore, are the statistics of elastic and inelastic scattering events that determine the trajectory of the particle, the energy distribution and the trajectories of the secondary electrons, and the substrate and its role in producing undesirable backscatter. Knowledge of these interactions allows trade-offs to be made between incident particle energy (or wavelength) and sensitivity per incident particle.

A useful figure of merit (FM) that exploits the various trade-offs is the weighted product

$$\text{F.M.} = (\text{edge slope})^a (\text{sensitivity})^b (1/\text{defect density})^c, \qquad (7)$$

where a, b, and c represent any (power law) weightings involved in the trade-off. No quantitative formulation of this figure of merit has yet been given for any of the lithographies. For purposes of the figure of merit, an isolated feature of, say, 0.5 μm could be used as the basis for the individual criteria.

The variation in absorbed energy per lateral distance at the defined edge of a resist feature is represented by edge slope. For UV exposure, edge slope is a function of wavelength, coherence, and resist thickness (Section III); for x rays, it is primarily a function of geometrical considerations in the engineering design (Section V); for e beams, it has a strong direct dependence on electron energy and an inverse dependence on resist thickness and beam spot profile (Section IV.D). Sensitivity is a function of the absorption coefficient for UV and x rays (Section IV.C) and increases for longer wavelengths of x rays; sensitivity is inversely proportional to (nearly the square of) energy for e-beam exposure (Section IV.D). The defect density has a strong inverse dependence on resist thickness for all lithographies.

A quick examination of the previous comments shows that the trade-off interplay is much greater for UV and e beams than for x rays. This gives a strong advantage to x rays (assuming that high-quality masks can be made) in that each of the factors in its figure of merit can be individually optimized. Edge slope (as defined by energy absorption, not the particulars of the resist development) for x rays is virtually independent of resist thickness, so the resist may be made thick to reduce the defect count. The absorption coefficients are generally weak enough for the shorter wavelength x rays that the energy absorption per unit volume will be uniform throughout the exposed depth. Finally, with x rays there is virtually no reflection or backscatter of the primary photons. Secondary electron effects at the interface can produce overexposure only if the x-ray mask contrast is sufficiently low and the substrate is dense.

A more complete figure of merit would include capital costs and throughput (partially dependent on sensitivity), but these will not be considered in this chapter.

B. Ultraviolet Scattering and Energy Loss

There is no optical scattering of any significance, but reflection from the substrate will strongly affect pattern linewidths and profiles. Reflection from the substrate can produce clearly resolved standing-wave patterns with node-to-antinode separations of $\lambda/4n$, where n is the index of refraction (~ 1.6). Standing-wave effects lead to linewidth control problems, particularly at steps. These effects are reduced with the use of structures such as those in Fig. 1.

A broad absorption or a poorly reflecting substrate will smooth the standing-wave pattern, but UV absorption spectra are fairly resonant, and this in turn accents the standing waves. Only a small part of the lamp spectrum is effective, with the sensitivity of most resists peaking at 3000 to 4000 Å. To improve resolution by using wavelengths of 2000 to 3000 Å, the longer wavelengths may have to be filtered and the exposure times must be increased. PMMA does not require filtering but has a low sensitivity. New, more sensitive resists for the 2000–3000-Å range are being developed.

Since borosilicate glass is absorbing for $\lambda \lesssim 3500$ Å, optical masks for the 2000–3000-Å range must use fused silica or sapphire substrates. To provide sufficient contrast in the mask, the thickness of the Cr absorbing layer must be increased. There are no practical mask substrates (or lenses) for $\lambda < 1800$ Å.

C. X-Ray Scattering and Energy Loss

In interactions with solids, x-ray photons may lose energy by various mechanisms. For the relatively low photon energies used in x-ray lithography ($E \approx 0.25$–5.0 keV for $\lambda \approx 50$–2.5 Å) only the photoelectric effect is of significance.

Coherent (Thomson) scattering, in which the scattered radiation is isotropic and has the same wavelength as the source, does not have a significant effect on x-ray resolution. It is the mutual interference among Thomson scattering events that generally gives rise to diffraction patterns. The fraction of the beam flux that is coherently scattered out of the beam and later loses energy within the resist layer is small, however, compared to losses by direct absorption. For all wavelengths greater than 0.5 Å, x-ray attenuation is dominated by true absorption rather than scattering. The Compton effect, which treats the photon as a particle and the scattering as a collision, also has no effect on resolution since the Compton electrons have energies of tens of electron volts or less.

The lower the incident photon energy, the greater is the photoelectric effect, i.e., the absorption of the photon by the atom with the isotropic emission of a photoelectron. The free energy of the electron equals the x-ray photon energy minus the initial binding energy of the electron. The photon absorption spectrum is continuous (Fig. 4) with selective absorption occurring at wavelengths close to those for x-ray emission. This is a desirable feature since it is possible to synthesize resists that are compatible with specific x-ray sources.

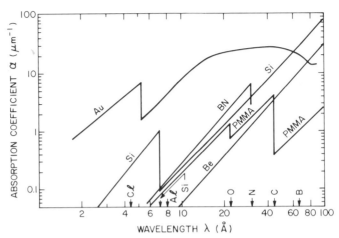

Fig. 4. Absorption coefficient versus wavelength for materials common to x-ray processing. Specific atomic absorption lines are indicated.

The absorption cross section for a photon of energy E by an atom of atomic number Z is $\propto Z^4/E^3 \propto Z^4\lambda^3$. Therefore, the inclusion of higher Z elements and the use of longer wavelengths greatly increases the absorption. The Z^4 dependence results from the involvement of the entire atom in the absorption. Conservation of momentum requires that momentum be transferred to the atom as a whole as well as to the emitted photoelectron. This is achieved via the Coulomb interaction between the electrons and the nucleus, with a stronger interaction and a stronger absorption being possible as Z is increased. The absorption of a higher-energy photon requires a larger transfer of momentum to the nucleus and produces the strong inverse dependence on energy.

The absorption, and therefore the energy loss, per unit distance is

$$dI/dx = -\mu I_0 \exp(-\mu x), \tag{8}$$

where I_0 is the initial beam intensity and μ the absorption coefficient. For $\mu x \ll 1$, $dI/dx \approx -\mu I_0$ and the energy loss per unit volume is constant. This condition is approximately satisfied for common resists for all wavelengths up to 15 Å (Fig. 4).

The photoelectrons lose energy in the same manner as electrons in e-beam lithography (Section IV.D) except that their rate of energy loss per unit distance ($\propto 1/E$) is much greater. The range of the photoelectrons in the resist determines the resolution unless the x-ray wavelength is so long that diffraction effects dominate. The optimum resolution occurs at the electron range–x-ray wavelength cross-over at 50 Å. For longer wavelengths, the photoelectrons have low energies and ranges shorter than 50 Å, and diffraction effects will determine the resolution.

Practical considerations in the design of the source and mask (Section V) favor shorter wavelengths. However, even for $\lambda = 5$ Å, the photoelectron range is only 600 Å in PMMA. For more dense resists, the photoelectron range is lower. The correlation of resist resolution with electron range has been experimentally confirmed. Features with resolutions less than 100 Å have been patterned.

The inherent resist resolution for x-ray lithography is excellent; better than 0.1 μm. Only ion-beam lithography offers the potential of better resist resolution. When the photon enters the resist, it is not scattered, and, when absorbed, it produces an intense, localized interaction with the resist by means of the photoelectron. There are no reflection problems and no serious diffraction problems. Vertical sidewalls of 1 μm in depth have been developed. Irregularities in x-ray mask patterns have been accurately reproduced; the x-ray resolution may, in fact, be limited by the mask resolution. Resist sensitivity may be improved by adding absorbers without degrading resolution. However, the combination of a high-sensitivity and a high-resolution resist is not presently available.

D. Electron Scattering and Energy Loss

For electron exposures the incident energy varies typically between 5 and 20 keV. At 5 keV the vertical range of the electron is not much greater than typical resist thicknesses and the lateral scatter is large. At 20 keV the energy loss per unit distance of travel within the resist layer is small but the lateral scatter is also small. Therefore, at 20 keV the sensitivity is lowered and the resolution is improved. Figure 5 shows iso-energy contours for 5- and 20-keV electrons incident on thick PMMA films (i.e., no substrate effects). Both contours are tear-shaped, but the contour for the 20-keV electrons is considerably more elongated.

The results of Fig. 5 can be easily explained qualitatively. For each scattering event between an incident electron and an atom of atomic number Z and atomic mass M, there is a Coulomb force (appropriately screened) between the electron and the nucleus. This force produces an impulse $F\tau$ with the force F proportional to Z and the interaction time τ inversely proportional to the electron velocity ($\propto E^{1/2}$). There is a momentum transfer to the nucleus proportional to $F\tau \propto Z/E^{1/2}$, which scatters the electron, and an energy transfer proportional to $(F\tau)^2/M \propto Z^2/EM$. There is a similar scattering and energy loss relation for each of the orbital electrons, i.e., with $Z = 1$ and $M = m$, the mass of an electron. The net energy loss to orbital electrons is therefore about 4000 times

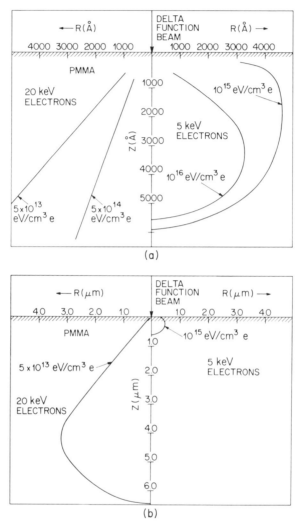

Fig. 5. Iso-energy contours for energy absorption within the PMMA resist when exposed to a point source of electrons. Absorption by 20-keV electrons is indicated on the left; absorption by 5-keV electrons is indicated on the right. Fig. 5b has an expanded z scale to indicate the tear-shaped absorption patterns and the greater range of the 20-keV electrons.

greater than the energy loss to the nuclei, with the energy loss per unit distance inversely proportional to E. Although energy loss is dominated by excitation and ionization of orbital electrons, the electron direction is dominated by the small-angle, nearly elastic scattering from the nuclei. In Fig. 5, the greater scattering for the 5-keV electron is obvious. The energy loss per unit distance of travel is also much greater for the 5-keV electron.

The energy loss per unit distance of travel occurs in a smooth, nearly continuous fashion, which can be represented by the Bethe [21] energy loss relation for nonrelativistic electrons,

$$dE/ds = -(2\pi e^4 n_e/E) \ln(1.166E/I), \tag{9}$$

where n_e is the density of atomic electrons and I the mean ionization energy for the atoms in the film. Everhart and Hoff [23] have normalized Eq. (9) and, by integrating over energy, have obtained universal curves for electron range versus energy. The range varies approximately as $E^{1.75}$.

Although a full spectrum of energy loss is possible at each interaction between the incident electron and an orbital electron, the probability of the incident electron losing a fraction of its energy $\Delta E/E$ while traveling a distance Δs is $\propto \Delta s/E^2$. Small energy losses are, therefore, greatly favored. The use of the continuous slowing down approximation (csda) of Eq. (9) avoids the complexity of incorporating the statistical probability for each energy loss and the need to track the paths of all secondaries. Calculations and experiments by Hawryluk *et al.* [27] have justified this approach.

The path of the primary electron can, therefore, be approximated by a series of straight-line segments between the elastic nuclear scattering events, with the Bethe energy loss expression being applied along each of the straight-line segments. The probabilistic nature of each scattering event must be handled by Monte Carlo calculations.

The theory of Spencer and Fano [22] provides corrections to the Bethe energy loss theory and reduces the estimate of the absorbed energy density for higher-energy electrons. Spencer and Fano take account of the discrete energy loss process rather than use the csda approach. A typical energy loss in one step for a 20-keV electron is ~ 100 eV. There are some modifications to predicted resist profiles when the Spencer–Fano theory is applied.

The probability of scattering through an angle ϕ during a distance of travel Δs in a material of atomic number Z is

$$p(\phi, E) = e^4 n_e Z \, \Delta s/16E^2 \sin^4(\phi/2). \tag{10}$$

This expression is derived by treating the atoms in the resist as individual scattering spheres. This is valid since the electron wavelength is itself so short (~ 0.1 Å). The fact that the incident kinetic energy of the electron is large compared to the binding energy of the atomic electrons allows the wave function of the system to be treated as the product of the individual wave functions plus a small perturbation representing the interaction. This Born approximation allows the asymptotic scattering to be accurately modeled. Both the differential and total (Rutherford) scattering cross sections vary as $1/E^2$. Small angle scattering is clearly favored. Since the scattering probability is proportional to Z^2/E^2, the scattering is greatly reduced by using higher-energy electrons and a low Z material.

In Monte Carlo calculations, the random statistics of the scattering are included, with appropriate weightings for the mix of Z values. The net result of a Monte Carlo calculation is the (average) energy dissipation per unit volume per incident electron as a function of the depth in the resist and the radial distance from the incident spot. The result is valid only for the particular incident electron energy, resist molecular configuration and thickness, and substrate for which it was calculated. A convolution integral may be used to convolve the delta function response of the Monte Carlo calculation with the actual beam profile to determine the net energy absorption per unit volume as a function of (x, y, z). Not only can "averages" be obtained but also standard deviations, such as straggling in the range of the incident electrons.

For PMMA resists, the number of elastic scattering events within a resist of thickness t is

$$N_e \approx 400t(\mu m)/E(keV). \tag{11}$$

For example, with $t = 0.4$ μm and $E = 20$ keV, $N_e \approx 8$. This gives a Gaussian standard deviation spread of ~ 400 Å to the initial delta function input. For $E = 10$ keV, $N_e \approx 16$ and the standard deviation of the scattering is ~ 1400 Å. For $E = 5$ keV, the N_e expression is no longer valid, but the scattering spread is $\gtrsim 0.4$ μm. Although sensitivity suffers significantly in employing energies greater than 10 keV, the resolution is also significantly improved.

A figure of merit, which included edge slope, was discussed in Section IV.A. For the practical example of a 0.5-μm isolated line, with its width defined by four Gaussian line exposures, the edge slope is increased by about a factor of 3 in going from 10 to 20 keV. For $E = 20$ keV, the edge slope is also increased by ~ 1.7 in going from a resist thickness of 0.5 to 0.25 μm. There are clearly strong trade-offs in e-beam exposure. However, resolutions of 0.1 μm can be achieved for thin resists and high ($\gtrsim 20$ keV) energies.

E. Ion Scattering and Energy Loss

Since the ultimate resolution in electron-beam lithography is limited by electron scattering effects and backscattering from the sample substrate, it is natural to pursue the applicability of ions as an exposure source. For the purposes of this discussion it is useful to consider ions in the energy range of 10 to a few hundred kiloelectron volts, although lithography schemes have been suggested that use still higher energy ions. This is understandable as scattering effects tend to decrease with increasing energy; however, practical engineering considerations tend to limit these possibilities.

The energy range from 10 to 300 keV is that associated with ion implantation and the basic energy loss and scattering processes are relatively well understood. Such ions lose energy primarily through inelastic excitations of the atomic electrons of the solid and nuclear recoil collisions with the atomic nuclei. They undergo angular deflections through scattering processes with the atomic nuclei. The intrinsic advantage of heavy ions over electron-beam systems is the relatively small lateral scattering of the heavy ion. This is most easily understood by considering the kinematic formula associated with the transformation of the center-of-mass scattering angle θ to the laboratory scattering angle Ψ:

$$\tan \Psi = \frac{\sin \theta}{(M_1/M_2) + \cos \theta} \tag{12}$$

where M_1 is the mass of the projectile and M_2 the mass of the scattering species. All values for the scattering angle θ are allowed. However, in the case where $M_1 \geq M_2$, there is a maximum scattering angle in the laboratory frame Ψ_{max}, given by

$$\sin \Psi_{max} = M_2/M_1, \qquad M_1 > M_2. \tag{13}$$

It is clear that heavy ions have a considerably smaller scattering angle for scattering from atomic nuclei and undergo essentially no deflections in scattering from electrons.

The rate of energy loss of heavy ions ($M > 10$) to electrons in this energy range $(dE/dX)_e$ is given quite accurately by

$$(dE/dx)_e = k\sqrt{E}, \tag{14}$$

while the nuclear contribution, of comparable magnitude, is a well-established but nonanalytical function, which decreases with increasing energy. The constant k can be calculated or, more accurately, extracted from experiment. As an example of the order of magnitudes involved, we note the following: the projected range of 100-keV Ga in an AZ111 photoresist (chemical composition $C_8H_{12}O$) is 0.22 μm with a lateral spread of 0.04 μm; the corresponding numbers for 500-keV Ga are 1.15 and 0.15 μm, respectively. The general requirements of very thin resists for lower-energy ions can lead to defect density problems during development.

In general, energetic ions are one to two orders of magnitude more efficient than 20-keV electrons for resist exposure. Qualitatively this can be understood in terms of the high charge of the heavy ion Z_1 and the velocity of the projectile; electron ionization cross sections tend to scale as Z_1^2 and maximize when the velocity of the projectile is close to the orbital velocity of the ejected electrons. This latter condition is more closely met for ions in the 100-keV range than for 20-keV electrons. In those cases where the exposure mechanism requires activation of two adjacent sites (such as in Novolac resists), the enhancement with heavy ions can be three or four orders of magnitude due to the higher density of the energy deposition track.

This enhanced sensitivity relaxes, to some extent, the current requirements in an ion-beam lithography system. This is not an insignificant consideration since the brightness of ion-beam systems tends to be small compared to electron-beam systems. Unfortunately, the real gain may be limited by statistical considerations. For example, the sensitivity of COP resists for 60-keV Ga has been reported as 10^{-8} C/cm^2 or an average of 10 ions/$(0.125\ \mu\mathrm{m})^2$ address element. Since the standard deviation in the exposure per address is $\sqrt{N}$, where N is the average number of incident particles, a large number of the elements would not be properly developed with such low dose (Section VI).

V. ENGINEERING CONSIDERATIONS

A. Engineering Considerations for UV Lithography

The optimum linewidth or space resolution is $\propto \lambda f$. The minimum depth of focus is $\propto \lambda f^2$. The image field, the area over which the high-resolution exposure can be achieved, is also a function of f. High-quality, large-image field lenses have larger f/numbers. For a full-field image diameter of 100 mm (4 in.), the lowest available f/number is about 3.5. With a λ of 3000 Å, λf is $\sim 1.0\ \mu$m. Submicron UV lithography can be achieved only for limited field diameters (and for reduced depths of focus).

For a field image diameter of 18 mm, the best available f/number is about 1.6 (at 4000 Å). This image field area corresponds to present LSI chip areas and, therefore, represents a reasonable lower bound on a step-and-repeat exposure system. If such an f/number can be achieved for a lens system with $\lambda = 2000$ Å, λf will be approximately 0.3 μm.

In order to cover the large field size of a 4-in. wafer, the field must be segmented by step-and-repeat (DSW) imaging or by scanning field projection.

In the DSW mode, a single chip pattern (or a small matrix of chip patterns) is used as the mask. The mask pattern is typically a 10× image and the optics provide a 10:1 reduction. The wafer is precisely stepped and exposure is made at each position. Realignment and refocusing are possible at each step. The best system currently available prints linewidths of $\lesssim 1$ μm and has a depth of focus of greater than ± 1 μm, with automatic focusing to ± 0.5 μm.

High-performance reduction lenses have traditionally been refractive. The required use of special glasses limits operation to the near UV. An advantage of the reduction is that it permits relaxed mask tolerances and geometries. However, the mask must be totally defect free, since a single defect on the mask will be stepped onto each chip of the wafer.

In the scanning mode, the mask is typically a full-field 1× mask. With the optics fixed, the mask and wafer are simultaneously scanned. A small optical field is permitted. The exposure is continuous and the scanning is at a uniform speed. The focus and registration must be maintained throughout the scan.

Scanning field projection permits the use of reflective optics, which has the advantage of operating in the deep UV with wider bandwidths. Alignment may be carried out with visible light and exposure with deep UV. The image field may even be smaller than the chip size, which gives the scanning system an edge as far as resolution is concerned. However, since there are currently no competitive reflective optics designs that work at magnifications of other than 1:1, the control of the mask dimensions is much more critical than for DSW printing with reduction exposure.

The resolution limits for deep UV should be ~ 0.3 μm. It is unlikely that production linewidths below 0.5 μm will be achieved within the next decade (Section VII).

B. Engineering Considerations for x-Ray Lithography

The scattering and energy loss processes associated with x-ray lithography offer resist resolutions of less than 0.1 μm. The resolutions that can actually be achieved for VLSI production are dominated by engineering considerations. An excellent summary of x-ray lithography has been given by Spiller and Feder [42].

Figure 6 is a schematic drawing of the Bell Laboratories' experimental x-ray lithography system. A 25-keV beam of electrons is incident upon a cone-shaped Pd target, which emits the characteristic L_α wavelength of 4.4 Å. The cone angle decreases the apparent source size (3 mm) at the wafer while allowing the incident electron-beam power to be absorbed over a larger area. The x-ray radiation passes through a beryllium vacuum

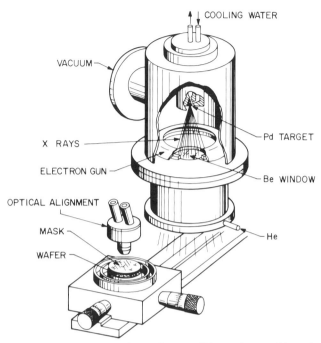

Fig. 6. Diagram of BTL's experimental x-ray lithography machine. A stationary, water-cooled, cone-shaped Pd anode is used as the source for the x rays.

window, in which the longer brehmsstrahlung wavelengths are absorbed, into a He-filled chamber to the mask and wafer. The characteristic L_α radiation (plus the shorter brehmsstrahlung wavelengths) is selectively transmitted through an Au-patterned polyimide-boron nitride mask to the resist-coated wafer. The space between the mask and wafer is flushed with He or N_2. The exposure time is about 0.5 min with 4 kW of source power.

Some of the major engineering considerations are the maximum power available from the source with reliable operation, the dimensional stability, the resolution and the contrast of the mask, and the various system dimensions such as source size, source-to-mask spacing, and mask-to-wafer spacing.

Practical considerations of the design of an x-ray source have been discussed by Maldonado *et al.* [52] and by Yoshimatsu and Kozaki [50]. In addition to the electron excitation of characteristic wavelengths, the use of continuous radiation from a synchrotron source has also been proposed. In a synchrotron, the continuous acceleration of relativistic electrons in a circular path results in the emission of broadband radiation, similar to brehmsstrahlung, which is highly collimated in the direction of electron motion and which has a small, wavelength-dependent divergence angle ($\sim 10^{-3}$ rad) in the direction normal to the plane of motion. If an exit port is placed tangentially to the circular path, the cone of radiation sweeps across the port to produce a narrow (3–4 mm), fan-shaped slit of intense radiation. The synchrotron avoids the severe intensity dropoff with distance of an isotropic, small-area source. Geometric distortion is also reduced if the total divergence angle is narrowed by inserting an absorbing filter for the longer wavelengths. On the other hand, higher mask contrast can be obtained with the inclusion of more exposure by soft x rays. Shorter wavelengths can be eliminated by a split-beam technique in which the portion of the beam reflected by a grazing incidence mirror is used for the exposure. The present disadvantage of the synchrotron is its cost.

The choice of a source with a characteristic line also involves several trade-offs. Figure 4 shows the absorption coefficient dependence on λ for several materials. The absorption coefficient (in reciprocal centimeters) varies between the discrete absorption edges as $\rho Z^4 \lambda^3$ (Section IV.C), where ρ is the density. (Appropriate ρ and Z weighting must be included for alloys or compounds.)

The most absorbing materials have an absorption coefficient that is only 50 times higher than the most transparent materials. The mask contrast, however, is determined only by the absorber. Longer wavelengths require thinner absorbing layers for a given contrast, so better mask resolution is possible. However, the mask substrate will also be more absorbing.

For a given vacuum window, mask substrate, and resist, it is possible to calculate the wavelength that will require the fewest photons at the source (i.e., the shortest exposure times). The less the absorption by the window and the mask substrate, the longer the optimum λ. The shortest exposure time should be possible in the range of the Si K line of 7.1 Å or the Al K line of 8.4 Å. Bell Laboratories has chosen the shorter Pd λ of 4.4 Å because the Pd line has a resonance match to the absorption by Cl in the BTL DCOPA resist, permits a thicker mask substrate, and can be used without attenuation in an He atmosphere. Br can be used for a resonance match to Si radiation.

The mask contrasts are somewhat better for Al and Si radiation than for Pd even though the absorption coefficient of Au has an absorption edge at 5.3 Å (Fig. 4), which tends to equalize the absorption coefficient at all three wavelengths. An Au absorber thickness of ~ 0.45 μm is necessary

for Si and Al to ensure a transmission contrast of 6; the thickness is increased to ~0.6 μm for Pd since the L line has a stronger short wavelength continuum. A thickness of 0.2 μm is required for the synchrotron
source with a 3-μm Si membrane (which absorbs the longer wavelengths
and centers the exposing radiation at ~10 Å). At the longer wavelength
of 44.7 Å for the C K line, the Au thickness must be only 0.07 μm.
Exposure to the C line must, however, take place in a vacuum.

A special advantage of x-ray exposure, and hard x rays in particular, is
the transparency of typical contamination particles such as dust or resist
that may accumulate on the mask (or wafer). The mask defect density
should not vary with time and use as is the case for UV masks.

An optimum mask substrate has not been established. In order to have
a 0.1-μm alignment accuracy over 10 cm, the dimensional stability must
be better than 1 ppm. Any random dimensional errors will translate
directly to the wafer. Organic films (Mylar, Kapton, etc.) are easy to fabricate but are not dimensionally stable. Inorganic films (silicon, silicon
oxide, silicon nitride, boron nitride, etc.) are difficult to fabricate and frag-
ile but are inherently more stable. Tensile inorganic films of 10 cm in
diameter can, nevertheless, be fabricated. The mask substrate must also
match to the source; e.g., membranes containing Si cannot be used with
Pd because of an absorption edge.

Figure 7 shows the geometric factors that affect the resolution. A pe-

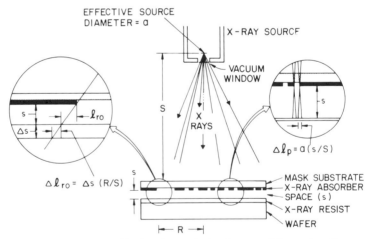

Fig. 7. Geometric factors affecting the resolution for x-ray lithography. The penumbral
blurring from the finite source size and the runout due to the mask-to-wafer spacing are indicated.

numbral effect results because of the finite source size a and the finite mask-to-wafer spacing s. The exposure time decreases as a is increased, but the penumbral blurring.

$$\Delta l_p = a(s/S),\qquad(15)$$

where S is the source-to-mask spacing, increases. The blurring of the absorbed energy at a feature edge results in a loss of resolution in the subsequent processing. The apparent source brightness can be multiplied and a decreased by using an oblique emission angle from an elongated spot, as in Fig. 6. However, oblique emission can introduce uniformity problems for large wafers (i.e., with a large angle subtended to the source).

Poorly defined edges of the mask patterns will also produce blurring. A radially dependent blurring also occurs from the partial transmission of the slanted rays through the shadowing edges of the mask patterns; this blurring scales linearly with the radial distance and inversely with the absorption coefficient.

A magnification from the mask to the wafer will occur because of the finite spacing s and the nonvertical incidence of the x-ray flux. The maximum magnification runout for a wafer of radius R is

$$l_{ro} = s(R/S).\qquad(16)$$

This magnification of the projected pattern can be used to compensate for uniform distortions of the wafer introduced by the individual, high-temperature, processing steps. The mask-to-wafer spacing s is a desirable parameter for providing this compensation.

Nevertheless, the runout can produce a misalignment from level to level if s is not precisely set to its proper value; i.e.,

$$\Delta l_{ro} = \Delta s(R/S).\qquad(17)$$

(For a synchrotron, the divergence angle of the source is much less than R/S for a typical "point" source.)

If requirements of Δl_{ro}, $\Delta l_p \leq 0.1\ \mu$m are set for a 10-cm wafer with a source $a = 3$ mm and a spacing $s = 10\ \mu$m, then S must be no less than 30 cm (to satisfy the Δl_p constraint) and Δs must be no greater than 0.6 μm (to satisfy the Δl_{ro} constraint). The Δs constraint is the more severe. It is possible to reduce the effective R by employing a reduced exposure field and a step-and-repeat exposure. Automatic "focusing" will also reduce the spread in Δs in the same way as for DSW UV printing. Exposure costs and exposure times will increase; however, not only will alignment be improved but mask stability problems will also be reduced.

C. Engineering Considerations for e-Beam Lithography

1. Electron-Beam Column Design–Mask Making

The limitations that arise in the patterning accuracy of an electron-beam mask-maker occur, not because of fundamental principles, but because of the need to write masks at an economic rate. By writing extremely small masks, the position errors can be made negligible, and by writing with a very small beam current, the spot diameter can be made as small as desired. However, in practice, a mask is required to have a large area containing many chips, and the beam current must be sufficient to expose the resist in a reasonable time.

First, let us consider the problems involved in writing a mask of large area (≥ 100 cm^2). A mask-maker differs from a direct writer in that registration marks are not available on the mask. The electron-beam machine must maintain its accuracy over the entire mask instead of over some subfield. The position of the stage can be monitored to within $\frac{1}{64}$ μm using laser interferometry (Fig. 8), and the column must then be able to write small subfields that must butt against each other very accurately. The main obstacle lies in the variation in the height of the mask, which can be caused by errors in the table motion or in curvature of the mask itself.

When the spot is translated by a deflection system in the electron-beam column, the beam will strike the writing surface at an angle β to the optical axis. If the writing surface is vertically displaced by a distance z, the spot will move laterally by a distance $\delta = \beta z$. If the maximum allowed value of z is limited to ± 15 μm (requiring good table design) and the maximum permitted position error is ± 0.05 μm, then β must be kept less than 3 mrad. This requires either a very long working distance or the use of a telecentric deflection system (which ensures that the beam always lands vertically, so $\beta = 0$). It is also theoretically possible to measure the height of the writing surface (optically perhaps) and to correct the deflection system gain, but this may be more complicated.

The variations in the mask height also give rise to errors in the spot size. Close to the best focus, the diameter d of the spot is greater than the minimum diameter d_0 according to the relation

$$d^2 = d_0{}^2 + 4\alpha^2 z^2, \qquad (18)$$

where α is the beam convergence semiangle and z the height error. If the spot is allowed to grow from 0.20 to 0.25 μm with a maximum height error of ± 15 μm, then $\alpha \leq 5$ mrad. Once again, if this constraint is unacceptable, direct monitoring of the mask height would be required.

Exposing a pattern on a mask when using a given electron resist requires the deposition of a fixed amount of electric charge. The shorter the time allowed to write the pattern, the higher the beam current must be. Let us examine the difficulties involved in obtaining a large beam current.

There are two general strategies that can be used to write a mask. The pattern can be built up by using a spot with a diameter approximately equal to the address size (0.25 μm perhaps). The spots are then exposed one by one until the whole mask is covered. In order to achieve high speed, the exposure time per spot must be very short and the current into the spot must be very high. On the Bell Laboratories' EBES-2 system (Fig. 8), the values are 50 nsec and 23 nA for a $\frac{1}{4}$-μm spot. For a $\frac{1}{8}$-μm machine, to write a mask in the same time would require a spot exposure time of 12 nsec and a current density 16 times higher. To obtain such current densities, a high brightness source (such as a thermal-field electron gun with a cathode of lanthanum boride, tungsten, or a tungsten alloy) must be used. These allow an array with 100-cm² area and with 25% coverage to be written in about 30 min (ignoring overheads), with an electron resist sensitivity of 10^{-6} C/cm².

An alternative technique, which avoids these difficulties and writes at longer spot exposure times and lower current densities, uses a greatly enlarged spot area. A larger spot with well-defined edges can be formed by illuminating a square aperture and then demagnifying this to form a square spot on the writing plane. The IBM El-1 system uses a spot 2.5-μm square with edges 0.4 μm wide and a beam current of 3 μA. If a spot is used that covers an area of 100 address points, then the electron gun source brightness and spot exposure repetition rate can both be 100 times lower than in the small spot scheme.

However, since many patterns cannot be made out of large squares and overlap causes unacceptable exposure variation in most applications, a spot of variable dimensions is required. This can be made by using two square apertures. The first is illuminated and imaged onto the second with an offset that can be changed with a set of deflection plates. Thus only a part of the second square will be illuminated and the size of this part is controlled by the deflection plate voltages. This illuminated area of variable size is demagnified as before to form the writing spot, and in this way the writing spot may be a rectangle with sides of any dimensions up to some maximum value. Most patterns may now be written without overlap, but difficulties with slanted lines remain.

Other schemes may achieve some of the advantages of both approaches. For example, two spot sizes could be used, one for outlining features and the other for filling in the interiors.

A major limitation in the variable-aperture scheme is imposed by space-charge effects. As the spot area is changed, the beam current will

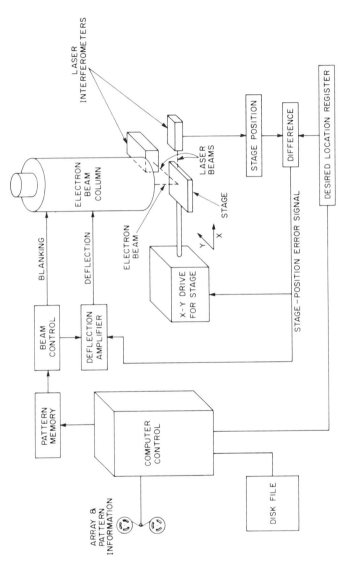

Fig. 8. Block diagram of BTL's EBES-2 electron-beam exposure system. The system employs a Gaussian beam spot in a raster scanning mode coupled with a continuously moving substrate table.

change (to preserve constant current density). This will change the total space charge within the electron beam, and at large currents this will change the plane of best-spot focus. This effect could be offset by using a dynamic focus coil to compensate for the defocus, but once again implementation would not be easy.

All schemes suffer from electron–electron scattering. When the beam current is substantially increased (to above 100 nA), the electron density will become so great that the electrons will be repelled from each other by coulomb interaction and will be scattered. This will give rise to an increase in the spot size even when the lenses are adjusted for best focus. The contribution to the spot size is approximately

$$\delta = 10^4 \, IL/V_0^{3/2}\alpha, \tag{19}$$

where I is the beam current, L the entire optical length of the column (i.e., the distance between the source and the mask), V_0 the beam accelerating voltage, and α the final-lens convergence semiangle (all in MKS units). For example, if $\alpha = 5$ mrad, $V_0 = 20$ kV, $I = 500$ nA, and $L = 30$ cm, then $\delta = 0.16 \, \mu$m. This imposes a severe limitation on the future development of fast, high-resolution, electron-beam lithography machines.

In the case of the imaged-aperture technique, this scattering effect reduces the edge sharpness of the larger spot in a similar manner. At 20 kV and 500 nA, the mean axial spacing of the electrons is 27 μm. Since the beam is much wider than this throughout most of the length of a column, the mean separation of the electrons will be much greater than this axial spacing. The scattering will thus be similar for all spot sizes up to at least 20 μm and will depend only on the total current. For a given required edge sharpness, an increase in spot area will not allow any increase in spot current.

2. Scanning Electron-Beam Systems

Most of the engineering limitations for e-beam lithography are cost-related, e.g., initial capital costs, operating costs, and rates of throughput. Broers [11] gives a starting point for cost comparisons among the various lithographies. The general advantages and disadvantages of e-beam lithography are briefly summarized in the following paragraphs. Resolution limitations are more specifically addressed.

Figure 8 gives a block diagram of the EBES-2 design of Bell Laboratories, which uses a Gaussian beam spot in a raster scanning mode with a continuously moving substrate table. Other systems use a variable-aperture spot, and/or a vector mode and/or a step-and-repeat table operation. Scanning of the beam is fully automated and software controlled.

The scheme used for beam patterning has a great effect on the length of time it takes to write a full pattern. The single-beam-spot mode is the slowest but the simplest from a data handling point of view. The variable-aperture mode is the fastest, but it is complicated by the design of the beam column and the data handling.

There are nearly 10^{12} 0.1×0.1-μm address spots in a 4-in. wafer exposure. A 0.1-μm-diameter beam spot in a full raster scan mode of operation with as little as 1 nsec of dwell time on each spot would require nearly 15 min of writing time. Such a short dwell time is compatible with presently attainable source brightnesses of $> 5 \times 10^6$ A/cm^2 sr and resists with sensitivities of $\sim 10^{-6}$ C/cm^2. However, present resists with sensitivities of $\sim 10^{-6}$ C/cm^2 do not approach resolution capabilities of 0.1 μm. With a 0.5-μm-diameter spot and a resist sensitivity of 10^{-5} C/cm^2, dwell times of as much as 80 nsec might be required to avoid spot size broadening of 0.1 μm from electron–electron interactions within the column [Eq. (19)]. Even with a vector mode of operation, the writing time would be about 15–20 min. Competitive production-line lithography systems should have writing times of less than 2 min. A system that combined a variable-aperture vector mode for the coarse features and a spot vector mode for the high-resolution features or for edge definition might be the only practical production system with 0.1-μm resolution.

If writing times approaching 2 min can be realized, then electron-beam direct writing will become a strong contender for submicron lithography. It is mask-free and, therefore, mask defect-free. It has an inherent means of providing registration by analyzing the backscattered electrons from the surface topography. With only three registration masks on a 3-in. wafer, alignment accuracies of better than 0.25 μm are presently being achieved across an entire slice. With a periodic reregistration step-and-repeat mode of operation, alignment accuracies of better than 0.1 μm can be achieved. Possible level-to-level misalignment produced by wafer distortions in x, y, or z are directly compensated by the software. Perhaps the strongest advantage of direct writing with e beams is its high accuracy of registration and its ability to compensate for local (step-and-repeat distance) and/or global distortions of the wafer. This capability justifies the use of e-beam exposure for levels of processing that require precise alignment. More and more attention will be paid to mix-and-match modes of lithography and to corrections for the various incompatibilities among lithography machines.

Electron-beam writing has a well-established learning curve since it has been and will continue to be used for making masks for UV and x-ray exposure. In fact, the x-ray and the UV 1:1 exposure patterns can be no better than the masks provided by the electron beam.

The resolution of e-beam exposure is limited by scattering within the resist and by backscattering from the substrate. The choice of resist thickness and beam energy can give resolution limits of better than 0.1 μm due to (forward) scattering within the resist. There may be less than ten scattering events for a 20-keV electron traversing 0.4 μm of resist. For practical structures with many microns of substrate thickness, the major limitation to the resolution is the backscatter from the substrate.

If the backscatter simply provided a uniform background level in the unexposed areas, analogous to background levels from the finite contrasts in masks, the backscatter exposure would be more tractable. The back-scatter dose, however, exhibits a "proximity" effect in which the absorbed energy in an unexposed area depends on the incident exposure to adjacent areas. Narrow gaps ($\lesssim 0.5$ μm) between large exposed pads ($\gtrsim 3.0$ μm) may not be resolvable. The proximity range increases as the incident electron energy increases. An electron of high incident energy undergoes very little scattering in initially traversing the resist; however, it enters the substrate with most of its energy remaining, penetrates deeply, and, therefore, backscatters over a wide range of several microns. The optimum trade-off occurs for incident energies of ~ 20 keV.

Proximity effects also occur within patterns. With uniform exposure, the net dose at the center of a small, isolated feature may be less than half that at the center of a large pattern and may, in fact, be less than the net dose in a narrow gap between large pads. High-contrast resists provide no solution in the latter case. Even if the feature sizes and separations are such that the maximum dose to any unexposed area is less than the minimum dose to any exposed area, the resolved features may still show an unacceptably large variation in the final feature widths as compared to the coded feature widths.

Partial solutions to the proximity problem are the judicious choice of a negative or positive resist depending on the actual pattern, design limitations on allowed pattern width and gap combinations, and the use of a multilevel low-Z substrate (Fig. 1). However, the point-by-point, or aperture-feature-by-aperture-feature, mode of exposure previously described as a disadvantage now becomes an advantage since the exposure at each address location can be individually adjusted. Since the focus may change if the beam current changes, the proximity effect corrections can best be made by modifying the dwell time via the deflection electronics. Large pads can, therefore, receive lower doses and isolated features can receive higher doses.

Since the dose at any address affects the dose at every other address, with a reciprocity relation between any pair of addresses, the solution as to what incident dose to apply requires either a self-consistent matrix formulation for pattern layout or an empirical algorithm. Therefore, software plays a paramount role both in preparing patterns for e-beam exposure

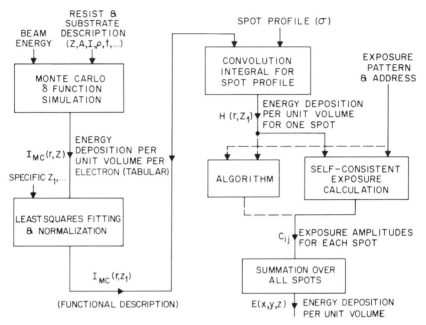

Fig. 9. Schematic diagram indicating the sequence of steps from an initial Monte Carlo simulation for a particular substrate to the final calculation of exposure intensities for each address location in the pattern to be written.

and in control of the beam itself. Figure 9 shows the sequence of steps from the original Monte Carlo simulation to the final definition of a pattern exposure.

3. Projection Electron-Beam Systems

There is one practical electron exposure system that makes use of masks and a flood electron exposure. This system is the photocathode projection (ELIPS) system, which is listed in the bibliography and will be discussed only briefly here.

The photocathode projection system uses a quartz mask with pattern features containing a photoemissive material. Ultraviolet light into the back of the mask is absorbed in the photoemissive material. The electron emissions are not necessarily normal to the surface but are nevertheless focused by uniform and parallel electric (E) and magnetic (B) fields to a corresponding point on the resist-coated silicon wafer which is parallel to the quartz mask. Some of the engineering difficulties in this system are: the precision required in the absolute value and uniformity of the E and B fields, the flatness requirements on the silicon wafer, the need to align the mask and wafer without unwanted exposure, the decay in sensitivity of the photoemissive material, the inability to correct for wafer distortions, and the inability to correct for proximity effect (other than by modifying the feature widths on the mask). Present systems can, however, achieve resolutions of about 0.5 μm and do offer a high throughput.

D. Engineering Consideration for Ion-Beam Lithography

1. Flood Exposure–Mask Design

The intrinsic advantages of heavy ion lithography discussed in Section IV.D can, in principle, be realized in two ways: (1) through the use of a focused ion beam for direct writing in a manner similar to the scanning electron beam and (2) through the use of an ion-beam mask in close proximity to a resist-covered substrate. In this section we shall concentrate on the mask technique; focused ion beams will be discussed in the following section. It should be realized that both of these ion-beam technologies are at an early stage of development and practical experience is limited.

Three types of masks have been considered for ion-beam lithography: open-stencil; thin, amorphous; and thin, single-crystal. Open-stencil masks do not use a supporting membrane and, hence, have essentially no deleterious effects on the transmitted beam. The resolution and resist profile under these conditions are optimal. However, since this type of mask cannot define patterns that close on themselves, the system is impractical.

A mask with absorbing features supported on a thin, amorphous membrane is a more general solution (Fig. 10). The range of energetic ions is a

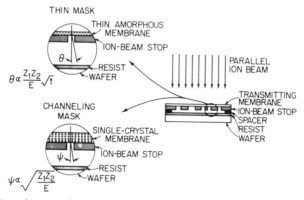

Fig. 10. A mask system for ion-beam exposure. The divergence angles for scattering by a thin, amorphous membrane or by a single-crystal membrane with channeling are indicated.

well-defined quantity and membranes can be constructed that transmit essentially 100% of the beam in the open areas. However, multiple scattering of the beam is an ever-present problem, which limits the membrane thickness and mask-to-wafer spacing. Multiple scattering is minimized by using as thin a membrane as possible; films of 1000 Å of aluminum oxide or boron nitride are close to the present limit. Under these conditions, the resolution due to 200-keV H beam spreading can be expected to be 0.1–0.2 μm, assuming a 20-μm mask-to-wafer distance. Multiple-scattering effects can be reduced by increasing the energy of the beam; however, a corresponding increase in feature height is necessary to stop the beam in the opaque regions. The mask design then approaches aspect ratio limitations, i.e., features of 1-μm widths and 5 μm-heights.

An elegant solution to the problem of multiple scattering is to use a silicon single crystal as a membrane and take advantage of the channeling effect (Fig. 10). Channeling occurs when an ion beam is incident on a single crystal parallel to a major symmetry direction. Under these conditions approximately 95–98% of the beam undergoes a series of correlated scatterings with the rows of atomic nuclei of the crystal, resulting in gently oscillating trajectories between the strings of atoms. Normal multiple-scattering processes are turned off and the emergent angular distribution is characterized by the channeling critical angle $(2Z_1Z_2e^2d/E)^{1/2}$, where Z_1 and Z_2 are the atomic numbers of projectile and crystal, respectively, d the atomic spacing along the atomic row, and E the incident energy. High-quality, single-crystal films as thin as 0.1 μm have been fabricated and easily withstand the damage that accompanies transmission of energetic ion beams. To first order the emergent angle is independent of membrane thickness, allowing the use of thicker membranes than in the amorphous membrane case. The channeling scheme might easily provide resolution limitations at the 0.1-μm or lower limit.

The limited experience with these techniques results in a number of unanswered questions requiring further investigation: (1) the effects of slit scattering associated with relatively tall and narrow features; (2) the ability to form structured masks on thin, Si, single crystals without deterioration of the crystal quality; (3) wafer or substrate damage; and (4) the actual formation of large, parallel, ion beams, mechanisms for step-and-repeat, registration, etc.

2. Ion-Beam Columns – Direct Writing

An ion-beam system would be similar in its general layout to an electron beam in a lithography system. The details of the components of the ion-beam column are under active consideration at the time of this writing.

Most conventional sources, such as are used in ion accelerators or sputtering stations, have too low a brightness and too large a chromatic spread for direct-writing, ion-beam systems. For example, a duoplasmatron source has a maximum brightness of 2000 A/cm^2 sr as compared with thermal electron sources with brightnesses of 10^7 A/cm^2 sr. Recently, however, two new types of sources, the field ion sources and the liquid-metal source, have produced brightnesses 10^4–10^6 times greater than the duoplasmatron and thus are good candidates for lithography systems. Experimenters have succeeded in demonstrating 400-Å lateral resolution with such sources for micromachining purposes.

While the absolute spatial resolution of an ion-beam system appears encouraging, the current requirements for direct writing with fine resolution still appear to have fundamental limitations. For example, to write a 4-in. wafer in the order of 15 min requires an address (0.125 μm × 0.125 μm) writing time of the order of 1 nsec and, for statistical accuracy, an exposure of 400 ions/address, implying a current of 64 nA. This current density requirement is limited (unachievable?) due to space-charge effects.

Space-charge blowup occurs as a result of the charges within an unneutralized ion or electron beam creating an electric field that disperses the charges in the plane perpendicular to the beam axis; i.e., the beam will expand in diameter. For a given mass, space-charge limitations tend to scale as $I/V^{3/2}$, where I is the current and V the kinetic energy of the beam. Wilson and Brewer [65] have supplied some useful tables and formulas to estimate space-charge limitations; in the previous example, the space charge would appear to limit the spot size to currents of the order of 1 nA for ions of mass 30 and kinetic energy of 100 keV. Neutralization of space-charge effects is possible and has been used in ion implantation systems. The secondary electrons produced in ion–atom collisions (with residual gas or walls of the vacuum system) or the electrons intentionally introduced into the system are trapped in the positive space-charge well of the beam, reducing the deleterious space-charge effects. On the other hand, neutralization of ion beams via electron pickup can cause difficulties; downstream focusing and deflection elements will have no effect on the neutral component of the beam.

Fundamental considerations of the motion of particles in electromagnetic fields suggest that the most convenient way to focus and deflect few hundred kiloelectron volt ions is with electrostatic elements. Electrostatic lens and deflection design is not as advanced as in the magnetic case, primarily due to the extensive use of the latter in electron microscopes, etc. Since the high resolution and small spot size of ion beams have already been demonstrated, the major consideration is deflected of the beam without substantial defocusing. Defocusing effects tend to be more severe in the electrostatic case (than in the magnetic case) since electrostatic elements tend to change the magnitude of the velocity of the ion. This restriction has limited the size of the scanned field to ~1 mm² in the most advanced prototype system built to date. Areas are then stitched together to cover the entire wafer through a registration mark system.

VI. STATISTICAL CONSIDERATIONS

Resist sensitivities are usually stated in terms of incident coulombs or joules per square centimeter. A more descriptive notation is the required number per unit area of incident particles (photons) of a particular energy for a resist of a stated thickness. This time-integrated flux, or count per unit area, is the necessary base for a statistical analysis.

Under the shot noise assumption that the individual particles arrive independently of one another, the probability of any particular number n of particles arriving in any specified area element is given by the Poisson distribution

$$f(n, m) = (m^n/n!) \exp[-m] \approx [1/(2\pi m)^{1/2}] \exp[-(n - m)^2/2m], \quad (20)$$

where m is the mean number of particles for the specified area element and $\sqrt{m}$ the standard deviation. Let a^2 be the resolution area, e.g., the address area for an e-beam exposure.

The statistical problem is to determine the probability that an exposed area receives a dose $n < n_{crit}$, where n_{crit} is the minimum dose necessary to resolve the element, and the probability that an unexposed area receives a dose $n > n_{crit}$. The value of m will, of course, differ for the two areas, with m ideally equal to 0 for the unexposed areas.

As an example, a probability of underexposure of 10^{-6} and a probability of overexposure of 10^{-7} for each area a^2 corresponds to an anticipated defect count of $(0.3 \times 10^{-6} + 0.7 \times 10^{-7}) (A/a^2)$ for a chip of area A, which has 30% of its area exposed. If $A = 1$ cm² and $a = 0.125$ μm, then the anticipated defect count is greater than 2000. If each such defect is fatal, then, clearly, the chip yield approaches zero. The probabilities of fatal underexposure or overexposure must be kept below $(1 - 10)(a^2/A)$ in order to have a practical yield.

Underexposure of an exposed area can occur with more likelihood if m is small, i.e., if the resist sensitivity is high. For any given resist, m decreases as a decreases, and the probability of underexposure *per element* increases. The important parameter is the ratio $K = |m - n_{crit}|/m^{1/2}$. The ratio K should be 6 or higher for submicron VLSI. It is for this reason that poor sensitivity, i.e., high m and high n_{crit}, is desirable for high resolution.

It is not acceptable to simply increase m by using a higher incident dose than is necessary for resolution without considering the effects of a higher dose on the probability for overexposure in the unexposed areas.

Direct-writing, ion-beam lithography is the only lithography that allows m to be increased with relatively no effect on the linewidth, other than that due to the edge profile of the beam itself, since the average flux of particles is effectively 0 in the unexposed areas and the cumulative lateral scattering within the resist is slight. The high sensitivity of resists to ion-beam exposure will, in fact, require that the incident dose be greater than that required for resolution.

Any lithography with masks will have an m for the unexposed (opaque) areas that scales linearly with that for the exposed areas. Increasing m in the transparent areas will decrease the probability of underexposure but will increase the probability of overexposure in the opaque areas. For x-ray exposure, m can be optimized to give the same probability for overexposure as for underexposure.

Let T represent the transparency of the opaque area relative to the transparent area. Therefore, the mean exposure in the opaque area is Tm. The probability of underexposure equals the probability of overexposure when

$$\sum_{n=0}^{n_{crit}} f(n, m) = \sum_{n=n_{crit}}^{\infty} f(n, Tm). \tag{21}$$

The approximate solution is

$$K_1 = (m - n_{crit})/m^{1/2} = K_2 = (n_{crit} - Tm)/(Tm)^{1/2}. \tag{22}$$

An n_{crit} of 30, with a T of 0.2, should ideally have an m of 67. The probability of overexposure or underexposure is $\sim 3 \times 10^{-6}$. Such a resist, in combination with a T of 0.2, is *too* sensitive for submicron VLSI. However, if T is reduced to 0.1, the ideal m is 95 and the probability of overexposure or underexposure is $\sim 1.4 \times 10^{-11}$. If n_{crit} is increased to 60, with a T of 0.2, the ideal m is 134 and the probability of overexposure or underexposure is $\sim 8 \times 10^{-11}$. The strong dependence on n_{crit} and T is apparent.

With an ideal step (with distance) from zero to full exposure, there is less edge growth as m is increased for x-ray exposure than there is with e-beam exposure. The cumulative effect of lateral scatter and backscatter is greater for e beams. That is, the edge contrast is determined not only by the exposure and development conditions for the resist but also by the spreading within the resist.

Although a direct-writing e beam does not use a mask, there is absorption in the unexposed areas by means of the electron backscatter from the exposed areas (as well as from the tail of the Gaussian beam). Increasing the beam current to increase m in the exposed areas will linearly increase the backscatter. Rather than having two separate defect probabilities as in the earlier examples, *each* exposed area has its own software-controlled value of incident m that depends on the expected contributions from its incident and backscatter exposure and the cumulative backscatter from neighboring elements. *Each* unexposed area receives a variable backscatter exposure that depends on the particular geometry of the neighboring patterns. Theoretically, the software control can be adjusted to give the same net absorbed dose in the exposed areas; however, a wide range of absorbed doses is possible in the unexposed areas. The absorbed energy in an unexposed area element can approach half (or more) of that in an exposed area.

The statistics of electron counting should separately include the incident electrons and the backscatter electrons from neighboring elements. The larger, effective, electron count reduces the probability of underexposure. The probability of overexposure varies inversely with the distance from a feature edge, so the unexposed area of concern is much less than for an x-ray mask exposure. The effective value of m for an unexposed area element can be represented as

$$m_u = \sum_i m_i P(r_i)(\alpha_i/2\pi), \tag{23}$$

where the summation is over all exposed elements, each of which receives a dose m_i, α_i the angle subtended by the unexposed element to the exposed element, and $P(r_i)$ the probability of any one electron being backscattered into the resist over the distance range to the exposed element. The standard deviation is not simply $m_u^{1/2}$ but rather

$$\sigma_u = \left[\sum_i m_i P^2(r_i)(\alpha_i/2\pi)^2 \right]^{1/2}. \tag{24}$$

The tighter distribution on m_u greatly reduces the statistical probability of overexposure.

With Tm in the earlier expressions replaced by m_u, σ_u becomes the standard deviation for the unexposed area elements. Thus, n_{crit} is lower for the unexposed areas than for the exposed areas since the energy deposited by a backscatter electron is greater than for the incident electron. A detailed solution has not been formulated.

A low-sensitivity resist, such as PMMA, which requires a large incident electron dose, will greatly reduce the statistical variation in the same way as in the examples for x-ray exposure.

VII. PROJECTIONS

Each of the lithographies discussed in this chapter is capable of submicron linewidths. Each is limited either by the exposing machine itself or by diffraction or scattering effects within the resist. The fundamental limitations due to molecular sizes within a polymer resist are on the order of 100 Å. The diffraction limits for UV exposure are ~3000 Å and for x-ray exposure are ~50–100 Å. The scattering limits, which are dependent on resist thickness for e beams and ion beams, are ~50–800 Å for x rays, ~200–3000 Å for e beams (with the upper limit set by worst-case conditions for proximity effect backscattering), and ~50–400 Å for ion beams.

For any specified line sizes there are also specifications on the dimensional control of the linewidths and spacings. Typical requirements might be ±10 to ±25% of the nominal value. Linewidth accuracy is, to a great extent, controlled by the inherent contrast of the resist, the development process, and the sensitivity to changes in the processing conditions. The resist properties may, therefore, dominate over the inherent lithographic properties in determining the final dimensional accuracy or practical resolution of the final patterns.

The introduction of linewidths below 0.5 μm into a production mode will require the solution to many problems, not the least of which is the development of high-contrast, reasonably sensitive resists. Dry-processed resists, whose linewidths and profiles are not modified by the swelling and flowing that occur in wet-developed polymer resists, should provide the linewidth control required in a production process. As the resist thicknesses are reduced, more attention will be paid to reduction in defect densities by going to finer filtering for the spin-on resists, or to evaporation, or to plating techniques. Higher-contrast resists will most likely come at the expense of lower sensitivity; this will also delay the introduction of submicron features because of the cost associated with longer exposure times. Poorer sensitivity is, however, a required corollary of finer resolution in order to reduce the statistical probability of exposure-related defects (Section VI).

The alignment control must be ~±25% of the minimum linewidth. Direct-step-on-wafer techniques for all of the major lithographies offer registration accuracies of ~±0.1 μm. For the near future these registration accuracies should be sufficient. The wafer distortions from process level to process level may, in fact, require DSW lithography for all mask–exposure systems in order to achieve the required alignment within the device patterns. Improvements in alignment accuracy can significantly reduce the total layout area for VLSI chips. For uniform magnification distortion, the mask-to-wafer spacing may be adjusted, without DSW, for exposure from "point" source x rays.

There are also control specifications on focus setting and mask-to-wafer spacing for each of the lithographies. Ultraviolet exposure is severely limited by its narrow depth of focus and will require both automatic alignment and automatic focusing in a DSW mode for submicron definitions. Mask-to-wafer spacing is crucial for x-ray exposure but can be controlled by capacitance sensing techniques. Although there are depth-of-focus limitations for direct writing by e beams or ion beams, they are not particularly restrictive.

Future lithography will be aided by adoption of the following trends in processing: high-contrast, dry-processed resists; more uniform wafer surfaces (e.g., multilevel resist structures); self-aligned device structures; lower temperature processing of device wafers; mix-and-match modes of lithography. Optimum costs may be achieved by using UV exposure for as many levels as possible, x-ray or ion-beam exposure for those levels requiring precise linewidth control, and e-beam exposure for those levels requiring precise alignment. The primary difficulty in such a scheme is to have matched mask sets and matched registration capabilities.

Figure 11 is an attempt to predict the future evolution of production-line lithography for structures with several levels of processing.

The Noyce [71] projections for linewidth reductions have been fairly accurate over the last decade, with production linewidths halving approximately every five years. If this trend is to continue, 1-μm linewidths will not be routinely available before 1985, and 0.5-μm linewidths will not be available before 1990. It is difficult to predict whether this trend will be enhanced or retarded. Retarding this trend are the increasing costs of the lithography processing, the resolution limitations of optical/UV exposure, resolution limitations of present resists, linewidth control, and defect density control. Enhancing this trend are the introduction of new processing techniques such as anisotropic, reactive sputter etching, the intensification of development efforts to match the demands of more complex LSI circuits, and the introduction of lithographies with inherently finer resolution capabilities. Direct writing will not be used extensively until the needs for its high sophistication justify its cost; its advent will be shortened by the development of a practical system that will reduce the total writing time by its ability to write both coarse and fine features.

Ultraviolet exposure will continue to be used throughout the 1980s, with bread-and-butter production leveling off at 1.5 to 2.0 μm. Submicron linewidths will not become common on a multiple-level process, production-line basis until at least 1986.

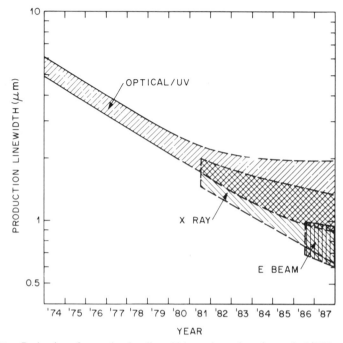

Fig. 11.   Projections for production linewidths and spacings for optical/UV, x-ray, and e-beam lithographies. The trends shown are for multiple-level production processing.

The first impact of x rays will not be in high-complexity VLSI circuits but rather in lower-complexity circuits for which higher speeds are desired. As yields improve and throughput increases, in the usual "learning curve" fashion, the lower costs with x-ray exposure should allow x rays to supplant many possible UV applications. The acceptance of x rays will depend crucially on alignment capability and, more specifically, on the dimensional stability of the mask substrate. The development of a high-resolution, dry-processed resist with a reasonable sensitivity ($<10^{-5}$ C/cm^2) will advance the application of x rays. Ultimately it may still be necessary to go to DSW in order to maintain alignments (and runout). A synchrotron source with multiple ports may have its cost justified for large-scale production applications.

Electron-beam lithography, as indicated in Fig. 11, will not accepted in a production-line mode until approximately 1986. Currently e beams are being used for some production, but, in those cases where the throughput is reasonable, the linewidths are no better than those that can be resolved by UV, and, in those cases where the linewidths are better than for UV, the throughput is on the order of one 3–4-in. wafer/hr. The 1986 date assumes the development of fast-deflection circuitry with 1-nsec spot exposures, sophisticated systems which permit both broad brush aperture exposure of patterns (or their interiors) and fine brush vector exposures of the perimeters of selected features, and enhanced software capabilities for proximity corrections. Electron beams should find their first applications in the via window level of integrated circuits. Proximity effect is not a problem for the isolated window features; the window is the most critical level for alignment and for determining layout area, and the total writing area is small enough to guarantee a high throughput. The e-beam registration must be compatible with the UV or x-ray registration for the other levels, but this problem has already been solved on individual systems. For multilevel applications, as implied in Fig. 11, e beams will initially find applications in special high-speed circuits and will continue to be used, as at the present time, for fast turnaround, low-quantity production.

It is clear that there are several solutions for the realization of submicron lithographies. The pressure for realizing submicron devices may really be on the controls available in the device processing itself.

ACKNOWLEDGMENTS

We are particularly grateful to L. C. Feldman for contributing the sections on ion beams and M. G. R. Thomson for contributing the section on the limitations of electron-beam columns. We are also grateful to R. K. Watts, J. H. Bruning, D. Maydan, M. Feldman, A. Zacharias, and G. N. Taylor for many discussions and for their helpful critiques.

REFERENCES

General Background (Resists, review articles, comparisons of lithographies, etc.)
1. L. F. Thompson and R. E. Kerwin, Polymer resist systems for photo- and electron lithography, *Ann. Rev. Mater. Sci.* **6,** 267–301 (1976).
2. M. J. Bowden, Electron irradiation of polymers and its application to resists for electron-beam lithography, *CRC Critical Rev. Solid State Mater. Sci.* **8,** 223–264 (1978).
3. M. J. Bowden and L. F. Thompson, Resist materials for fine line lithography, *Solid State Technol.* **22(5),** 72–82 (1979).
4. G. N. Taylor, X-ray resist materials, *Solid State Technol.* **23(5),** 73–80 (1980).
5. G. N. Taylor and T. M. Wolf, Plasma developed x-ray resists, *J. Electrochem. Soc.* **127,** 2665–2674 (1980).
6. S. Yamamoto, K. Kobayashi, and Y. Toyama, Application of electron beam resist PMMA to semiconductor device fabrication, *Fujitsu Sci. Technol. J.* **14,** 143–157 (1978).
7. A. Barrand, C. Rosilio, and A. Ruaudel-Teifier, Monomolecular resists: A new class of

high resolution resists for electron beam microlithography, *Solid State Technol.* **22(8)**, 120–124 (1979).

8. J. M. Moran and D. Maydan, High resolution, steep profile resist patterns, *Bell Syst. Tech. J.* **58**, 1027–1036 (1979).

9. M. Hatzakis, D. Hofer, and T. H. P. Chang, New hybrid (e-beam/x-ray) exposure technique for high aspect ratio microstructure fabrication, *J. Vac. Sci. Technol.* **16**, 1631–1634 (1979).

10. R. K. Watts, Advanced lithography, *in* "Very Large Scale Integration (VLSI)— Fundamentals and Applications" (D. F. Barbe, ed.), pp. 42–88. Springer-Verlag, Berlin, 1980.

11. A. N. Broers, High resolution lithography systems, *Physikalische Blatter* **34**, 704–712 (1978).

12. G. R. Brewer, A review of electron and ion beams for microelectronic applications, *Proc. Int. Conf. Electron Ion Beam Sci. Technol., 4th* 455–480 (1970).

13. Y. Tarui, Basic technology for VLSI, *IEEE Trans. Electron Devices* **ED-27**, 1321–1331 (1980).

Diffraction, Energy Loss, and Scattering

14. M. Born and E. Wolf, "Principles of Optics," 5th ed., pp. 459–532. Pergamon, Oxford, 1975.

15. J. W. Goodman, "Introduction to Fourier Optics." McGraw-Hill, New York, 1968.

16. A. Walther, *in* "Applied Optics and Optical Engineering" (R. Kingslake, ed.), Vol. 1, p. 245. Academic Press, New York, 1965.

17. M. Kakudo and N. Kasai, "X-Ray Diffraction by Polymers." Kodansha, Tokyo, 1972.

18. A. E. Hughes and D. Pooley, "Real Solids and Radiation," pp. 98–123. Wykeham, London, 1975.

19. Chr. Lehmann, "Interaction of Radiation with Solids and Elementary Defect Production," pp. 1–150. North-Holland Publ., Amsterdam, 1977.

20. H. S. W. Massey and E. H. S. Burhop, *in* "Electronic and Ionic Impact Phenomena" (W. Marshall and D. H. Wilkinson, eds.), Vols. 1 and 2. Oxford Univ. Press, London and New York, 1969.

21. H. A. Bethe, Moliere's theory of multiple scattering, *Phys. Rev.* **89**, 1256–1266 (1953).

22. L. V. Spencer and U. Fano, Energy spectrum resulting from electron slowing down, *Phys. Rev.* **93**, 1172–1181 (1954).

23. T. E. Everhart and P. H. Hoff, Determination of kilovolt electron energy dissipation vs. penetration distance in solid materials, *J. Appl. Phys.* **42**, 5837–5846 (1971).

24. J. S. Greeneich and T. Van Duzer, An approximate formula for electron energy vs. path length, *IEEE Trans. Electron Devices* **ED-20**, 598–600 (1973).

25. R. D. Heidenreich, L. F. Thompson, E. D. Feit, and C. M. Melliar-Smith, Fundamental aspects of electron beam lithography. 1. Depth-dose response of polymeric electron beam resist, *J. Appl. Phys.* **44**, 4039–4047 (1973).

Proximity Effects and Monte Carlo Calculations

26. T. H. P. Chang, Proximity effect in electron beam lithography, *J. Vac. Sci. Technol.* **12**, 1271–1275 (1975).

27. R. J. Hawryluk, A. M. Hawryluk, and H. I. Smith, Energy dissipation in a thin polymer film by electron beam scattering, *J. Appl. Phys.* **45**, 2551–2566 (1974).

28. M. Parikh, Self-consistent proximity effect correction technique for resist exposure (SPECTRE), *J. Vac. Sci. Technol.* **15**, 931–933 (1978).

29. M. Parikh and D. F. Kyser, Energy deposition functions in electron resist films on substrates, *J. Appl. Phys.* **50**, 1104–111 (1979).

30. K. Murata, E. Nomura, K. Nagami, T. Kato, and H. Nakata, A three-dimensional study of the absorbed energy density in electron-resist films on substrates, *Jpn. J. Appl. Phys.* **17**, 1851–1860 (1978).

31. K. Murata, M. Kotera, and K. Nagami, Remarks on the calculation of energy loss in electron-resist films on substrates, *Jpn. J. Appl. Phys.* **17**, 1671–1672 (1978).

32. E. Nomura, K. Murata, and K. Nagami, Fundamental studies of the interproximity effect in electron-beam lithography, *Jpn. J. Appl. Phys.* **18**, 1353–1360 (1979).

33. N. Sugiyama, K. Saitoh, and K. Shimizu, Proximity effect correction in EB lithography for VLSI microfabrication, *Digest IEEE ISSCC* 88–89 (1979).

34. J. S. Greeneich, Impact of electron scattering on linewidth control in electron-beam lithography, *J. Vac. Sci. Technol.* **16**, 1749–1753 (1979).

35. H. Sewell, Control of pattern dimensions in electron lithography, *J. Vac. Sci. Technol.* **15**, 927–930 (1978).

Optical/UV

36. *SPIE Dev. Semicond. Microlithogr. I, II, III, IV, V, SPIE Conf. Proc.* **80, 100, 135, 174, 221** (1976, 1977, 1978, 1979, 1980).

37. J. D. Cuthbert, Optical projection printing, *Solid State Technol.* **20**(8), 59–69 (1977).

38. D. A. Doane, Optical lithography in the 1-μm limit, *Solid State Technology* **23**(8), 101–114 (1980).

39. J. W. Bossung and E. S. Muraski, Optical advances in projection photolithography, *SPIE Dev. Semicond. Microlithogr. III* **135**, 16–23 (1978).

40. G. L. Resor and A. C. Tobey, The role of direct step-on-the-wafer in microlithography strategy for the 80's, *Solid State Technol.* **22**(8), 101–108 (1979).

41. M. J. Buzawa and A. R. Phillips, Ultraviolet objectives for submicron photolithography, *SPIE Dev. Semicond. Microlithogr. III* **135**, 77–82 (1978).

X Ray

42. E. Spiller and R. Feder, X-ray lithography, *in* "X-Ray Optics—Applications to Solids" (H. J. Queisser, ed.), pp. 35–92. Springer-Verlag, Berlin and New York, 1977.

43. S. Nakayama, T. Hayaska, and S. Yanazaki, X-ray lithography, *Rev. Elec. Comm. Lab.* **27**, 105–115 (1979).

44. R. Feder, E. Spiller, and J. Topalian, Replication of 0.1 μm geometries with x-ray lithography, *J. Vac. Sci. Technol.* **12**, 1332–1335 (1975).

45. D. C. Flanders, Replication of 175Å lines and spaces in polymethylmethacrylate using x-ray lithography, *Appl. Phys. Lett.* **36**, 93–96 (1980).

46. E. Spiller, D. E. Eastman, R. Feder, W. D. Grobman, W. Gudat, and J. Topalian, Application of synchrotron radiation to x-ray lithography, *J. Appl. Phys.* **47**, 5450–5459 (1976).

47. S. Matsui, K. Moriwaki, S. Hasegawa, H. Aritome, and S. Namba, Contrast of the x-ray mask for synchrotron radiation and the characteristics of replicated pattern, *Jpn. J. Appl. Phys.* **18**, 1205–1206 (1979).

48. T. Nishimura, H. Kotani, S. Matsui, O. Nakagawa, H. Aritome, and S. Namba, X-ray replication of masks by synchrotron radiation of INS-ES, *Jpn. J. Appl. Phys. Suppl. 17-1* **17**, 13–17 (1979).

49. K. D. Kolwicz and M. S. Chang, Silver halide–chalcogenide glass inorganic resists for x-ray lithography, *J. Electrochem. Soc.* **126**, 135–138 (1980).

50. M. Yoshimatsu and S. Kozaki, High brilliance x-ray sources, *in* "X-Ray Optics—Applications to Solids" (H. J. Quesser, ed.), pp. 9–23. Springer-Verlag, Berlin and New York, 1977.

51. D. Maydan, G. A. Coquin, H. J. Levinstein, A. K. Sinha, and D. N. K. Wang, Boron nitride mask structure for x-ray lithography, *J. Vac. Sci. Technol.* **16**, 1959–1961 (1979).

52. J. R. Maldonado, M. E. Poulsen, T. E. Saunders, F. Vratny, and A. Zacharias, X-ray lithography source using a stationary solid Pd target, *J. Vac. Sci. Technol.* **16**, 1942–1945 (1979).

E-beam

53. A. N. Broers and R. B. Laibowitz, High resolution electron beam lithography and applications to superconducting devices, *Am. Inst. Phys. Conf. Proc.* **44**, 289–297 (1978).
54. J. P. Scott, Electron-image projector, *Philips Tech. Rev.* **37(11–12)**, 347–356 (1977).
55. J. P. Scott, Recent progress on the electron image projector, *J. Vac. Sci. Technol.* **15**, 1016–1021 (1978).
56. W. D. Grobman *et al.*, 1 μm MOSFET VLSI technology: Part VI—Electron beam lithography, *IEEE Trans. Electron Devices* **ED-26**, 360–368 (1979).
57. E. V. Weber and R. D. Moore, E-beam exposure for semiconductor device lithography, *Solid State Technol.* **22(5)**, 61–67 (1979).
58. D. R. Herriott, R. J. Collier, D. S. Alles, and J. W. Stafford, EBES: A practical electron lithographic system, *IEEE Trans. Electron Devices* **ED-22**, 385–392 (1975).
59. P. R. Thornton, Electron physics in device microfabrication I, *Adv. Electron. Electron Phys.* **48**, 271–380 (1979).
60. J. L. Mauer, H. C. Pfeiffer, and W. Stickel, Electron optics of an electron-beam lithographic system, *IBM J. Res. Dev.* **21**, 514–521 (1977).
61. M. G. R. Thomson, R. J. Collier, and D. R. Herriott, Double-aperture method of producing variably shaped writing spots for electron lithography, *J. Vac. Sci. Technol.* **15**, 891–895 (1978).
62. A. V. Crewe, Some space charge effects in electron probe devices, *Optik* **52**, 337–346 (1978/79).
63. K. Hoh, N. Sugiyama, and Y. Tarui, Optimum beam size for high-speed drawing of LSI patterns by variably shaped rectangular electron beams, *IEEE Trans. Electron Devices* **ED-26**, 1363–1365 (1979).
64. W. Knauer, Analysis of energy broadening in electron and ion beams, *Optics* **54**, 211–234 (1979).

Ion Beam

65. R. G. Wilson and G. R. Brewer, "Ion Beams." Wiley, New York, 1973.
66. N. Deornaley, J. H. Freeman, R. S. Nelson, and J. Stephen, "Ion Implantation." North Holland Publ., Amsterdam, 1973.
67. R. L. Seliger, R. L. Kubena, R. D. Olney, J. W. Ward, and V. Wang, High resolution, ion-beam processes for microstructure fabrication, *J. Vac. Sci. Technol.* **16**, 1610–1612 (1979).
68. R. R. Hart, C. L. Anderson, H. L. Dunlap, R. L. Seliger, and V. Wang, High current density Ga^+ implantations into Si, *Appl. Phys. Lett.* **35**, 865–867 (1979).
69. D. S. Gemmell, Channeling and related effects in the motion of charged particles through crystals, *Rev. Mod. Phys.* **46**, 129–227 (1974).
70. M. Komuro, N. Atoda, and H. Kawakatsu, Ion beam exposure of resist materials, *J. Electrochem. Soc.* **126**, 483–490 (1979).

Projections

71. R. N. Noyce, Large-scale integration: What is yet to come? *Science* **195**, 1102–1106 (1977).
72. W. T. Lynch, The reduction of LSI chip costs by optimizing the alignment yields, *Technical Digest—1977 IEDM Meeting*, 7G-7J (1977).
73. R. Allan, Semiconductors: Toeing the (microfine) line, *IEEE Spectrum* **14**, 34–40 (1977).

Chapter **4**

Research and Resource at the National Submicron Facility

E. D. WOLF
J. M. BALLANTYNE

National Research and Resource Facility for Submicron Structures
Cornell University, Ithaca, New York

I. INTRODUCTION

Microstructure science and engineering, as shown by the chapters in this book, is a diverse field, which draws its practitioners and their techniques from a wide range of disciplines. Such a field needs one or more national centers where research workers from different backgrounds can come together, where appropriate experimental equipment can be concentrated, and where information can be gathered to serve the nation's research community. As the field developed without such a center, a very noticeable gap opened between university research on the one hand and

129

the accomplishments of industrial laboratories on the other—a gap due mainly to the expensive equipment and the interdisciplinary nature of microstructure science and engineering that universities found difficult, if not impossible, to support. Sensing this widening gap, the National Science Foundation established a national facility for university research in this field in 1977. Originally housed in temporary quarters at Cornell University in Ithaca, New York, the facility will move its operations to a new laboratory on the Cornell Campus in 1981. One purpose of this "focused" facility is to stimulate university research by providing an equipment base for visiting scientists from other universities, as well as for resident Cornell scientists, who could not otherwise afford programs in microstructure research. The facility also serves as a resource for workers in government and industrial laboratories across the country. Its general objectives are

to foster research on methods for building submicron structures and to encourage expansion of the science base required for submicron engineering, as well as to educate in the field;

to provide a facility where research workers with different types of science and engineering background and from many different institutions can build experimental structures, devices, and systems needed in research that involves submicron feature size;

to establish a center of expertise in submicron structures that will serve as an information resource for the research community.

To help accomplish these objectives, the facility was established on an interdisciplinary faculty base at Cornell University and is governed by the Policy Board and Program Committee with major representation from outside the Cornell community. The first five-year NSF grant totaled $5.5 million. A major fraction of this was used in the initial capital equipment purchase, leaving an average annual budget for the remaining four years of the grant of about $850,000 per year. This annual budget supports three major categories: (1) central facility staff, (2) core research projects, which draw on the interdisciplinary faculty and student base at Cornell, and (3) capital equipment maintenance, upgrading, and new purchases. The central facility staff consists of the director, senior research associates, engineers, and technicians (see Section III.B).

Programs operating in the facility are expected to explore advanced areas that intimately depend on submicrometer dimensions and to leapfrog such areas as current integrated circuit practice. Individuals or institutions can interact with the facility in a variety of ways depending on the nature of the project. Ideal projects would meet all of the following criteria:

(a) Projects should involve microminiaturization, with emphasis on submicrometer dimensions in a substantial and innovative way.

(b)   The project should have goals that substantially further the art of submicron technology or its applications to engineering or scientific research. Such projects will probably be most successful when they involve either substantial collaborative interaction with a local member of the facility or substantial resident work at the facility by senior personnel from the proposing institution.

(c)   Projects should not be of a "service" nature where such services are commercially available.

(d)   Projects should be of such a nature that the specialized equipment or expertise in the facility makes an essential and important contribution to the outcome of the work.

(e)   Projects should provide educational opportunities for personnel associated with the work.

The facility should be viewed mainly as an equipment and technical resource rather than a financial one, since very few funds are available from the facility for direct support of user research projects. While no charges are presently levied for the use of equipment or facilities in the laboratories, users should, in general, expect to secure support from other sources for the operating costs of their projects, such as expendables, materials, travel, and personnel.

Interested individuals who wish to use the resources of the National Submicron Facility have a choice of ways to proceed. They can, after preliminary discussion with the director submit a short proposal outlining the objectives of their project and estimating what equipment would be used over what period of time. The Program Committee, meeting quarterly, reviews these proposals for acceptance or rejection. Such projects are necessarily supported by the user's own funding and require only the specialized equipment and expertise available at the facility.

Another way to interact with the National Submicron Facility is by way of NSF Joint University–Industry Cooperative Research Programs. The Cornell Program for Submicrometer Structures (PROSUS) offers more conventional interaction with industrial affiliates. Collaborative proposals to other funding agencies and exchange of personnel are other possible ways to become involved. Work done in the facility is subject to the rules of Cornell University, which generally require that it be nonproprietary and publishable.

The charter of the facility is very broad as far as the type of work that is deemed appropriate. Most programs fall into three categories: projects that advance the art of submicron fabrication technology, those that utilize the technology to fabricate devices or research structures with submicrometer dimensions, and those that address fundamental materials or important physical problems on the submicron scale. In order to ensure a broad approach to the field of microstructure science and engineering, the facility was established at Cornell with close research ties to the Schools of Electrical Engineering, Applied and Engineering Physics, and Chemical Engineering, and the Departments of Materials Science and Engineering, Chemistry, and Physics. Some projects involve representation from areas of the biological sciences, computer science, and astronomy. A flavor of the breadth of the research programs can be gained from the following section in which some of them are discussed in more detail.

II. RESEARCH PROGRAMS

Essential to a facility of this type are ongoing core research projects that establish a base of scientific expertise with which users and visiting scientists can interact. Fourteen projects of this kind involving Cornell personnel (typically a graduate student and/or a postdoctoral research associate on each project) were supported by the National Submicron Facility during the 1979–1980 fiscal year. In addition, some sixteen user/visitor projects were approved for access to the facility during the same period. These projects are listed in Tables I and II. These core and user research programs fall generally into the areas of lithography and pattern transfer; device design, fabrication, and characterization; materials research and fundamental studies; and circuit design and layout. While some programs overlap more than one of these categories, some representative projects are discussed under these headings in the following sections. It is impossible for a small facility such as NRRFSS to address all areas within its broad charter. Therefore, an effort has been made to choose core and user research topics that offer pivotal new science and technology opportunities. The overall strategy is based on the assumption that industrial research will push conventional devices toward their classical limits. Hence, this facility should emphasize the transition between classical device limits and new device physics below 0.1 μm in feature size. This does not mean that 0.1–1.0-μm devices and research structures are excluded. On the contrary, they are studied as basic physical, chemical, and electrical links or transitions to smaller devices. These studies are also our main link to industrial interests in the facility. However, the facility emphasis will be to extrapolate by experiment and theory to new, smaller devices and their understanding. This is an important criterion for choosing projects within the facility. This emphasis is apparent in the description of the ideal project given in the Introduction.

TABLE I

Core Research List 1979–1980

Researchers[a]	Projects
J. M. Ballantyne (EE)	Materials and Technology for Integrated Optical Devices
B. W. Batterman, D. Bilderback, and R. A. Buhrman (A&EP)	X-Ray Lithography and Microscopy Facility Associated with CHESS
R. A. Buhrman (A&EP)	Submicron Lithography and Ion Processing
L. F. Eastman and C. E. C. Wood (EE)	Selected Area Molecular-Beam Epitaxy
J. Frey (EE)	High-Speed SOS Device and Materials Studies
D. W. Hammerstrom (EE)	NRRFSS Computer System Research
M. Isaacson (A&EP)	Production and Characterization of Metal Lines Less than 30 Å in Width and Computer Resource for Electron Optics
C. A. Lee (EE)	Ion-Beam Lithography and the Exposure Characteristics of Resists Exposed with Ions
T. Rhodin (A&EP)	Physics of Compound Semiconductor Interfaces
A. L. Ruoff (MS&E)	Diamond x-ray Source
B. M. Siegel (A&EP)	Development of a High-Current, High-Resolution Field Ion Probe for Lithography and Archival Memory
J. Silcox (A&EP)	Exploratory STEM Studies
C. L. Tang (EE)	Integrated Optical Circuits and Devices
E. D. Wolf (EE)	Reactive Plasma Etching

[a] A&EP, Applied and Engineering Physics; EE, Electrical Engineering; MS&E, Materials Science and Engineering.

A. Lithography and Pattern Transfer

One consequence of the project emphasis as previously described is that lithography and processing efforts in the facility are directed toward those technologies that can be used successfully in defining or transferring patterns at the 0.1-μm level and below. Figures 1 and 2 identify the most promising types of lithography for this region, namely: x-ray lithography, electron-beam lithography, and ion-beam lithography. Since there is a large amount of industrial development of electron-beam lithography systems with resolution in the 0.5–2.0-μm area, this type of development work is not being emphasized in facility research programs. Ion-beam lithography possesses some substantial advantages for defining patterns with dimensions of the order of 0.1 μm or less. Ion lithography is much less developed than electron-beam lithography and, hence, is one area of

TABLE II

1979–1980 User/Visitor Projects at NRRFSS

Researchers	Projects
Joseph T. Boyd, Dept. of Electrical Engineering, Univ. of Cincinnati	Travel Expenses to Perform Lithography for V-Groove Channel Waveguide Fabrication
W. S. C. Chang, Dept. of Electrical Engineering, Univ. of California	Research and Development of Microfabrication Technology for $LiNbO_3$ and $LiTaO_3$
L. F. Eastman and C. E. C. Wood, School of Electrical Engineering, Cornell Univ.	Optimized Nucleation and Growth of Semiconductor Heterojunction Structures by Molecular-Beam Epitaxy for Improved Microwave Optoelectronic and Superconducting Devices
T. N. Rhodin, School of Applied and Engineering Physics, Cornell Univ.	Physics of Compound Semiconductor Interfaces — Semiconductor Heterojunction Structures Produced by Molecular-Beam Epitaxy
J. Osteryoung, Dept. of Chemistry, SUNY, Buffalo	Electrochemistry at Very Small Electrodes
V. Kapoor, Dept. of Electrical Engineering, Case Western Reserve Univ.	Imaging Charge Coupled Devices with Submicron Features
J. C. Beck, H. L. Hegdus, J. Belina, and R. Sacco (Supervising Faculty W. Heetderks), School of Electrical Engineering, Cornell Univ.	Design Project Proposal for Microelectrode Array
S. Shapiro, Dept. of Electrical Engineering, Univ. of Rochester	Submicron Josephson Effect Weak Links
M. Harwit, Dept. of Astronomy, Cornell Univ.	UV Polarizing Grid for Spacecraft Photo Polarimeter
R. Mattauch, Dept. of Electrical Engineering, Univ. of Virginia	Production of Molecular-Beam Epitaxial Layers to Be Used in Millimeter-Wave Mixer Research
R. Richardson, Dept. of Physics, Cornell Univ.	Submicron Structures at Very Low Temperatures
J. Reppy, Dept. of Physics, Cornell Univ.	Weak-Link and Josephson Structures for Superfluid Helium Experiments" (Scope Reduction)
M. A. Littlejohn, Dept. of Electrical Engineering, North Carolina State Univ.	Research on $Ga_{1-x}In_xP_{1-y}As_y$ and Its Application for Submicron Devices
Rishi Raj, Materials Science and Engineering Dept., Cornell Univ.	Submicron Grids on a Metal Surface
G. C. Dalman and C. A. Lee, Dept. of Electrical Engineering, Cornell Univ.	Submicron Millimeter Wave Diodes
P. Batson, IBM T. J. Watson Research Center	STEM Research

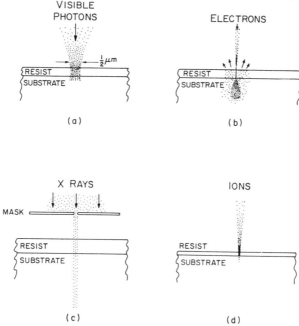

Fig. 1. Types of advanced lithographies for submicron structure fabrication. (a) Photolithography, the mainstay of the microelectronics industry, is used routinely down to 2 to 3-μm linewidths; the practical limit may approach 0.5 μm. (b) Electron-beam lithography, which has produced microstructures on thin films below 100 Å, suffers from backscatter electrons on bulk substrates when the pattern elements are closer than about 0.5 μm. (c) X-ray lithography is very effective to near 100 Å but requires a relatively thick absorber-mask pattern on a thin supporting membrane. (d) Ion-beam lithography offers a tremendous potential for very high resolution microstructure fabrication because of the large energy dissipation near the surface and the short range of secondary excitation processes.

program emphasis within the submicron facility. Work is going on in the development of bright ion sources and in the optics necessary to form and scan fine ion probes, as well as in the study of ion-beam exposure mechanisms of polymer and inorganic resists.

1. Bright Ion Source

A project under the direction of Professor Benjamin M. Siegel in the School of Applied and Engineering Physics is being carried out in conjunction with Dr. Gary R. Hanson to develop a gaseous field ion source with high angular current [1]. They have built and carried out initial tests on a field ion source that operates at very low temperatures with a physisorbed gas supply to the ionization region. A schematic of the system is

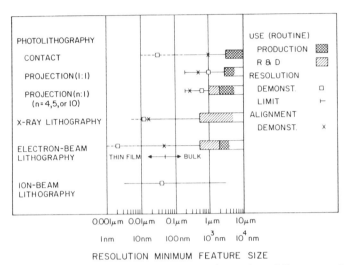

Fig. 2.   Approximate resolution (minimum feature size) and capability status of advanced lithographies.

shown in Fig. 3. The H_2^+ ion-beam currents of 2 to 20 nA into half-angles of 7 to 10 mrad have been measured, giving angular beam currents $dI/d\Omega \approx 10$ to $60\ \mu A/sr$. The measured energy spread is about 1 eV at the lower beam currents [2]. The UHV system allows processing of the field emitter under clean, ultra high vacuum conditions. A closed cycle, liquid He cooled, cold finger maintains the tip at controlled temperatures from that of liquid He to room temperature. The tip is mounted on a sapphire block to provide both excellent thermal conduction and electrical insulation of 30 kV. Differential pumping allows a high supply pressure of H_2, He, Ar, etc. in the region of the tip (10^{-2} torr) and a low pressure in the rest of the system. Observations on the characteristics of the field ionization pattern have been made under varying conditions of pressure, temperature, and field, as well as tip processing and orientation by viewing the field ion pattern on a channel electron multiplier array (CEMA) and phosphor screen. An alignable aperture defines the angular divergence. The H_2^+ beam appears to be very stable, giving no observable current fluctuations as long as the gas supply and tip temperature are kept constant. The low-frequency flicker noise usually associated with gaseous field ionization at liquid N_2 temperature is not observed with this source at liquid He temperature. At the current stage in the development of the H_2^+ field ion source, beams of 10 $\mu A/sr$ are routinely produced and, under special processing of the tip, beams of 60 $\mu A/sr$ are obtained.

A retarding field energy analyzer has been designed, fabricated, and installed in the field ion system and is used to measure the energy width of the ion beam under different conditions and current densities. The retarding field analyzer is of special design with five electrodes for retarding and reaccelerating the beam. The electrodes are of titanium and the insulators are Macor ceramic. The entire analyzer has been brazed into a single unit using an active metal braze method developed in this project by Hanson [3]. The method allows simple, active, metal brazing of Ti to ceramic materials and greatly simplifies the construction of electrostatic ion and electron optical systems.

Hanson and Siegel have also produced Ar^+ ion beams of high angular current and are currently investigating methods for optimizing the beam currents for the rare gases Ar^+, Kr^+, etc. The heavier ions would be used for direct sputter etching or irradiation-enhanced etching of the substrates and thin-film layers, which are used to fabricate different types of devices.

Work is underway to focus and deflect these beams with an electrostatic–ion-optical system to form probe diameters from 0.5 μm down to 10 nm and to investigate the interactions of the H_2^+ ions and other gaseous ions such as Ar^+, O_2^+, and N_2^+ with resists such as PMMA and the materials used in submicron devices and structures. As will be shown, the H_2^+ ion probe has great promise for ion-beam lithography at

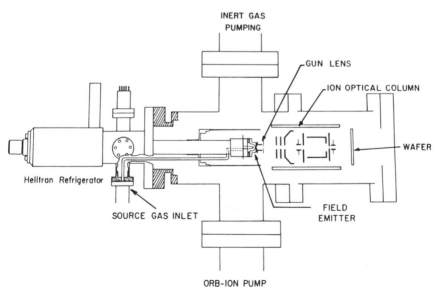

Fig. 3. Schematic of the field ion source with ion optical column for ion-beam lithography on wafers. The Helitron refrigerator operates at 0.7 liters per hour, using 0.5 W at 4°C.

high writing rates without proximity effects. The interaction and sputter etching of the heavier ions will be investigated to evaluate the possibilities of direct fabrication of structures and devices. Systematic scaling experiments will be made down into the nanometer range to study the special effects and characteristics of materials as the linear dimensions are reduced. Important new properties and opportunities for research are anticipated.

2. Ion-Beam Lithography

Professor C. A. Lee and Leo Karapiperis in the School of Electrical Engineering are studying the ion-beam exposure characteristics of poly(methyl methacrylate) (PMMA) and have demonstrated the potential of producing very narrow lines in PMMA on thick substrates by means of ion-beam exposure [4]. They exposed 2800 Å of PMMA to a dose of 2×10^{-6} C/cm^2 of 40-keV H$^+$ ions in an ion implanter through a conformal gold mask 1100 Å thick. The mask was fabricated using a laser holographic technique to produce a fine grating in the photoresist on top of the gold film, followed by ion milling to transfer the grating to the gold. Narrow windows down to less than 400 Å were produced. Lines as narrow as 400 Å with almost vertical walls were produced in the PMMA after H$^+$ bombardment through the mask and subsequent development (cf. Fig. 4).

Although this work demonstrated resolution capability at the few hundred angstrom level and greatly increased exposure sensitivity relative to electrons or x rays, a more quantitative comparison between theory and experiment is required to understand the exposure and development profiles at these resolution levels. One of the limitations of the gold mask fabrication technique was that the walls defining the slit for exposing the underlying PMMA were not sufficiently vertical. Thus the edge of the proton exposure was graded by the absorption in a wedge of gold. Various techniques have been tried to improve the mask definition. Changes in the electroplating and ion-milling techniques used in conjunction with the holographic process did not yield significant improvements. Encouraging results with mask linewidths down to 2000 Å have been obtained with lift-off techniques and electroplating used in combination with photolithographic masking.

A more promising technique is under development using the SEM and/or the STEM machine to produce narrow lines in PMMA spun on specially prepared thin membranes. The PMMA will be used in both its positive and negative modes to fabricate narrow linewidth gold masks. Quantitative exposure and development characteristics of high molecular

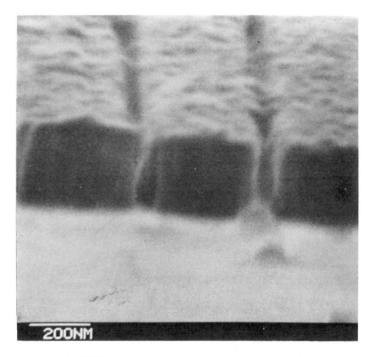

Fig. 4.   Grating in PMMA exposed by 40-keV protons through a gold conformal mask (not shown). Linewidths of 400 to 500 Å are evident.

weight PMMA will be studied. In collaboration with I. Adesida, a Monte Carlo simulation of ion-beam scattering and energy deposition in PMMA is being made on the basis of the LSS theory [5a–c], and the results will be correlated with experiment.

3. Carbon x-Ray Source

The technique of x-ray lithography has been shown to be very promising for replicating patterns with very high resolution and for achieving high-aspect ratios in patterns with dimensions less than 1.0 μm. One of the highest resolution resists is the conventional PMMA, for which it has been shown that carbon K radiation (λ = 44.8 Å) is an ideal wavelength for optimal resolution. Professor A. L. Ruoff and his associates in the Department of Materials Science and Engineering are involved in a program to construct a high-brightness carbon K radiation source for x-ray lithography purposes. Their approach uses diamond as the anode or target material. It is attractive mainly because of its high thermal conductivity and high strength. They predict [6] that a type IIb diamond of thickness

0.01 cm might sustain a power input of 6.5 kW while being bombarded with electrons (6 kV) on a stationary 1-mm-diameter area, and that this power input would generate carbon K x rays of sufficient brightness to expose PMMA in 1 min for a source to substrate distance of 2.5 cm.

They have designed and fabricated an efficient cooling system for the electron-beam, irradiated diamond target. With the present design, they have achieved a power input of 0.6 kW in an 0.8-mm-diameter area on a diamond of diameter 0.6 cm and thickness 0.045 cm without any visible damage to the diamond surface. Higher power input could not be achieved due to limitations of the electron gun and the charge buildup on the diamond surface (type IIb diamonds were unavailable at the time). They are now working with type IIb diamonds that are sufficiently large in size and have a low enough electrical resistivity to meet design criteria. It is, therefore, expected that with these diamonds and the present design they will soon have an operating system that will work at power inputs of nearly 2 kW on a 1-mm-diameter area. To achieve higher input powers, a new electron gun must be designed. Work is in progress toward this end, and it is anticipated that with the new electron gun and anode assembly power inputs of more than 4 kW will be achieved—not far from the goal set by the Nelson and Ruoff calculations.

4. Narrow-Line Technology

Several different techniques are being developed by Professor R. Buhrman and co-workers in the School of Applied and Engineering Physics for the purpose of establishing a useful lithographic capability at and below the 0.1-μm-linewidth level. Some of these techniques were previously developed by others and are being applied to specific problems in the facility. A major objective of Professor Buhrman's program on submicron lithography and ion processing is the production of high-aspect ratio patterns (greater than 4 to 1) at 100-nm linewidths and below.

One approach being taken is to produce master patterns in PMMA electron resists by operating the Cambridge 150 SEM in the line scan mode or with a flying spot scanner. By operating at a reasonably high accelerating voltage (30–40 kV) to reduce secondary electron broadening, simple patterns have been produced and replicated into gold by standard evaporation and lift-off procedures. The narrowest gold lines achieved to date have been 70 nm wide and 100 nm high. An example of such lines is shown in Fig. 5. These results appear to be near the limit achievable on solid substrates with single-layer resists. Using a two-layer resist process, whereby two resists of different exposure and development characteristics are used to alleviate the problems of broadening due to backscattered electrons and low-aspect ratios, 60-nm-minimum linewidths in

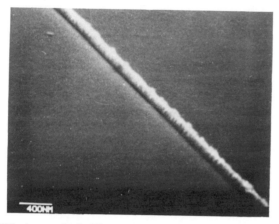

Fig. 5.   An SEM micrograph of a 70-nm gold line produced by electron-beam lithography.

two-layer resists of 250-nm total thickness were already achieved in pre-liminary experiments. This work is continuing.

Fine-line patterns have been replicated on a small-scale vacuum x-ray lithographic system. Gold absorber patterns were produced by photon and electron lithography on supported thin (1–2 μm) membranes of polyimide. Using these patterns as a mask, PMMA was exposed using the L x rays from a copper source. An example of a low-density, high-resolution pattern produced in PMMA by this method is shown in Fig. 6.

Reactive oxygen ion etching of polymer films is being developed as a

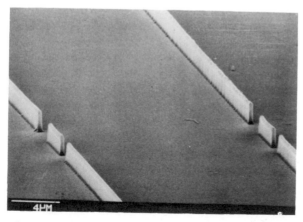

Fig. 6.   An electron-beam exposed test pattern replicated by x-ray lithography in PMMA resist.

Fig. 7. An edge pattern in polyimide produced by masking and reactive oxygen ion milling.

technique for producing high-aspect ratio patterns [7]. In the work to date, thin metal masks (Ti or W) are patterned on thick, 2–3000-nm, polymer films (polyimide or PMMA) by either photon or electron lithography. This pattern is then replicated into the polymer by ion milling with a mixture of oxygen and argon ions using the Commonwealth ion-beam system (the argon ions are simply to slow the process down). Very deep patterns with vertical walls can be readily obtained in this manner. The resulting polymer pattern then can be used as a standard lift-off stencil. An example of an oxygen ion etched edge in polyimide is shown in Fig. 7. These high-aspect ratio patterns are also being used to produce specialized masks for x-ray lithography following the technique of Flanders [8].

5. Resists and Processing

The development of new resists also comes under the purview of the facility in the area of lithography and pattern transfer. While our program in organic resists, which involves Professors Rodriguez from Chemical Engineering, Obendorf from Human Ecology, and Wolf from Electrical Engineering, is just getting underway, some work on inorganic resists has been going on slightly longer and has produced interesting results. Since inorganic resists exhibit good adhesion, mechanical and thermal stability, and since they are totally nonswelling and nondeforming, they are quite compatible with dry processing technology. Because they also have very high resolution, they are very interesting candidates for the fabrication of small structures where the total avoidance of wet processing is attractive.

Professors Ruoff and Chopra and Dr. Balasubramanyam are investigating the Ge_xSe_{1-x} resist system and have shown that the optical contrast increases markedly with the angle of deposition of the films. Effects of x-ray and proton beam irradiation on these films is being studied, as is the kinetics of dry etching techniques for obliquely deposited films [9].

Professor Edward Wolf is initiating a research program on reactive plasma etching. The emphasis in this program is on the fabrication of submicron structures in conjunction with electron-beam lithography and subsequent patterning of silicon, silicon-on-sapphire, and aluminum surfaces and on the etching and ashing of resists including double-layered structures. Future work is planned on plasma etching of III–V semiconductor compounds and refractory metals. Initial work is being carried out in a diode-planar reactor with a unique experimental research chamber supplied by Applied Materials, Inc. The reaction chamber has a symmetric configuration to facilitate etching at either electrode and to allow for variations in plasma excitation and grounding combinations. In some related work, reactive ion etching of molybdenum is being done by people in Professor Ballantyne's group for the purpose of preparing substrates for graphoepitaxy of GaAs. Although done in a different diode sputtering system, work to date has produced very sharp edges and smooth bottoms on grooves in molybdenum substrates.

6. Limits of Electron-Beam Lithography

The facility is sponsoring programs to develop the techniques of electron-beam lithography for producing lines less than 0.1 μm wide. This is a region that is well beyond the present industry thrust in developing electron-beam lithography for microcircuit applications. In this regime, electron-beam exposure must either be done on a thin membrane to avoid backscattered electrons or other novel techniques must be applied. One program in the facility under the direction of Professor Michael Isaacson in the School of Applied and Engineering Physics has the objective of producing and characterizing metal lines less than 3 nm in width. The program relies on the use of the scanning transmission electron microscope (STEM) to write and characterize the structures produced. While the STEM has been used previously to write lines less than 10 nm wide, the main drawbacks with the existing systems have been that the use of thermionic sources precluded high currents in subnanometer diameter probes (for instance, in Broers [10], only 10^{-12} A was available in a $\frac{1}{2}$-nm probe), resulting in inordinately long exposure times. Also, the existing systems did not include microchemical analysis capabilities to allow characterization of the structures produced. With the advent of field emission source technology with source brightness several orders of magnitude greater than existing thermionic sources, nanoampere beam currents become available in subnanometer diameter probes. It thus becomes reasonable to try to explore the ultimate limits of lithography since different materials may be studied that need not be as radiation sensitive as the standard resists being used, and researchers can begin to examine the real atomic and molecular limits.

The questions that arise are: (1) What physical processes limit the ultimate linewidth? (Can we ever hope to fabricate structures of atomic dimensions?) (2) What new methods can be derived for nanometer structures that are not necessarily extensions of the methods used to fabricate $0.1-1.0$-μm linewidths? (3) How can we characterize such lines (e.g., structurally, electrically, and chemically)?

These questions are being pursued using the Vacuum Generators Microscope HB5 STEM equipped with a cold field emission source. The system can operate reliably at 100 keV and can easily produce more than 1 nA of beam current in a 0.5-nm-diameter spot, 10 nA in a 5-nm spot, and 130 nA in a 50-nm spot. (See Fig. 8 for experimental measurements on the system at Cornell.) Thus, exposures of greater than 1 C/cm^2 in 10 μsec for nanometer-size probes are obtainable. The sample chamber can be maintained at $0.5-1 \times 10^{-9}$ torr vacuum and the hydrocarbon contamination level can be kept below any desired minimum value (e.g., a 1-nm-diameter probe with 1-nA current on a 20-Å-thick carbon film produces no evidence of contamination after irradiation times as long as $\frac{1}{2}$ hr). Furthermore, with the addition of a high-luminosity, high-energy-resolution electron spectrometer, which is now under construction ($\frac{1}{4}$-eV resolution out of 100 keV), the capability of generating and analyzing structures less than 3 nm in size will be in hand. This capability has been verified by preliminary results in a program on beam writing in NaCl films. An example of an ex-

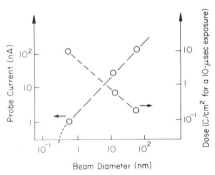

Fig. 8. Beam current versus spot size for the STEM for a total emission current less than 20 μA at 100 keV.

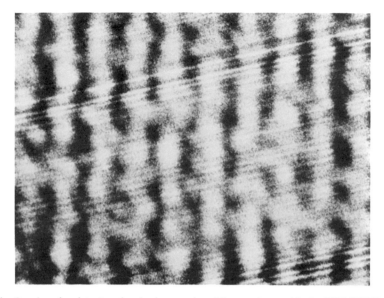

Fig. 9. Annular detector signal micrographs of lines written with the HB5 STEM on a sample consisting of a thin carbon substrate onto which ~ 1–4-nm-diameter In balls were deposited. About 20-nm thickness of NaCl was deposited over that. The indium balls appear white on the dark background, as does the unexposed NaCl. The full horizontal scale is 750 Å and the exposed dark lines are about 30 Å wide.

posed line written into a thin (~ 20 nm) NaCl film solvent deposited onto a thin carbon substrate (3-nm thick) is shown in Fig. 9. Small indium particles were evaporated onto the carbon substrate prior to the NaCl deposition, both to aid nucleation and create a more uniform thickness film and for measuring the probe size. Preliminary results indicate linewidths less than 3 nm. These are the narrowest lines ever reported for a resist. Production of wider lines (see Fig. 10) seems to indicate that the edge width (or edge resolution) is of the order of the beam diameter. At the time of this writing, an investigation of the parameters that control the line edge width was just beginning. The initial results would appear to indicate that if the vaporized trough can be filled with metal, line-aspect ratios greater than 5 to 1 can be realized.

Plans for characterizing such small structures include development of the following methods:

(1)   Using the annular, dark-field detector signal, very small mass thickness changes can be detected with minimum exposure [11] and this signal, coupled with the energy loss signal (ELS), can be utilized to determine the average atomic number [12]. The annular detector, which is

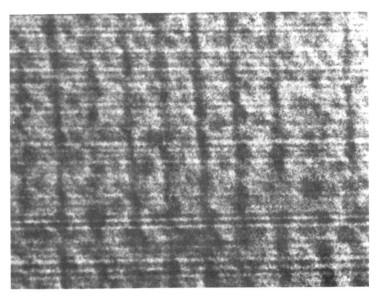

Fig. 10. Bright field signal micrographs of lines written in NaCl using the STEM. The full horizontal scale is 750 Å and the unexposed regions appear dark. The exposure is fifteen times that given in Fig. 9 and the white exposed lines are about 60 Å wide with edge resolution comparable to the spot size.

standard equipment on the VG HB5 STEM, has a poor detection efficiency (DQE 1:10). It will be replaced with one of our own design, which will have a DQE 1:2.

(2) Convergent beam diffraction techniques will be used to determine the substrate thickness to better than 1% accuracy for crystal substrates (see the work of J. Silcox in Section II.C). In order to optimize this technique, we shall ultimately require the capability of parallel recording of the diffraction pattern rather than the sequential method used now of sweeping the pattern across a small aperture and displaying the intensity on a CRT. Toward this end, an annular fluorescent screen will be installed to record the diffraction pattern simultaneously with viewing the bright field image (or while recording an energy loss or x-ray spectrum). Ultimately, a two-dimensional, charge-sensitive array for parallel recording of the diffraction pattern is desired.

(3) Because the standard HB5 electron energy loss spectrometer is quite crude, a high-performance spectrometer has been designed to replace it. Once installed, ELS will be utilized to determine the conductivity and the electronic structure of the small metal lines.

(4)   ELS and x-ray analysis will be used to characterize the line composition locally.

7. Electron Optics and Electronics Design

Because of the importance of electron- and ion-beam optics to a number of instruments in the facility, a library of programs for computer-aided design of electron optics has been assembled and is continually being expanded. These programs not only enable the electron optics of most current electron-probe-forming systems to be analyzed but also assist in the design of new components, such as the new electron-energy-loss spectrometer for the STEM previously mentioned, and the design of the optical column for the focused ion probe. Professor H. Ohiwa from Toyohashi University of Technology, Toyohashi, Japan, has collaborated with Professors Isaacson and Siegel in the establishment of this library and in the design of the optics for the ion probe system. One part of this library is the computer program MOL contributed by Professor Ohiwa, a brief description of which follows.

The computer program MOL was developed for designing magnetic e-beam scanning systems in a conversational or interactive mode with a computer. It calculates third-order deflection and first-order chromatic aberrations together with paraxial properties for a system consisting of round lenses and cosine-wound deflectors. These lenses and deflectors can be positioned at any points so that the lenses and deflectors can be superposed on each other and the deflectors can be twisted relative to each other.

For calculating magnetic fields, an analytic expression is used, known as the Grivet-Lenz model (the z-dependence of the lens field is expressed by the hyperbolic secant function). A similar model field is used for deflectors as well. Hence, the system simulated by this program must be expressed by a superposition of these assumed models. The advantages of using the analytic expressions are (1) the availability of higher-order derivatives of the fields, which allows use of a ray-tracing method for evaluating paraxial and aberration properties, and (2) the fast computation, which enables conversational or interactive usage of the computer.

Paraxial trajectories and aberrations are expressed in ordinary differential equation form, and they are evaluated by the Runge–Kutta method. By evaluating redundant trajectories and aberrations, computational errors can be estimated for the results.

In addition to the conventional optical properties of the simulated systems, aberrations on adding predeflection at the object point are also calculated. The predeflection was introduced with the concept of the moving objective lens, but the simulated system need not be a moving objective lens. Indeed, the predeflection often reduces deflection aberrations of other systems as well. Ways to apply predeflection are also calculated so that predeflection can be realized by using an actual deflector.

The program was designed to be used in a conversational or interactive mode. Two things are important for this: first, the results must be obtained within a few seconds; second, computer input for specifying the simulated system must be as simple as possible. The first point was realized by using the ray-tracing method and analytic expressions for magnetic fields. On a PDP 11/34 with a single user, the computation time for a system is several seconds. The second point was realized by coding the input in such a way that the user need input only the minimum necessary information, and this can be done in arbitrary order, using simple commands. These points enable the user to concentrate on his design procedure and free him from long delays in getting results and from tedious and error-prone data preparation.

In addition to work on design and electron optics, a modest program is underway to develop special-purpose, electron-beam pattern generators. The objectives of this work are to provide the SEM and STEM in the facility with modest digital pattern generation capability and to develop new pattern generators, which may be required for the fabrication of integrated optical structures. The approach taken was to regard the entire pattern generation system as composed of individual boxes driven from a common digital bus and having a common analog output, each box containing hardware pattern generation capability optimized for specific applications (such as rectangle fill-in, vector generation, raster generation, etc.). The digital input bus is obtained from the PDP-11 DR11B parallel input/output device. Initial work has been on a hardware rectangle generator (RG) and on a digitally controlled precision analog raster generator (ARG). The ARG will be useful in writing periodic structures for optical applications with much higher precision than can be achieved from D/A converters. The RG is designed for 16-bit resolution, stepping speed to 10 MHz, point-by-point plotting at a rate limited by the DR11B (between 50 and 100K points/(sec) and allows software specification of fill-in mode, clock rate (ECL clocks can be set to $\pm 0.2\%$ for rates between 10^{-3} and 10^7 Hz), and blank time so as to allow software control of proximity correction.

B. Device Design, Fabrication, and Characterization

Current programs in the area of device design and fabrication stress areas that benefit from very small dimensions and have substantial potential for future applications. Some are directed toward fundamental measurements rather than toward electronics.

1. High-Speed SOS

A core research program on high-speed, silicon-on-sapphire (SOS) device and material studies is under the direction of Professor Jeffrey Frey in the School of Electrical Engineering. The objective of this work is to explore the usefulness of submicron-scale silicon structures for very high-speed linear and digital applications. To achieve this objective, the following programs are being carried out:

(1)   a study of fabrication methods for low-parasitic, submicron-channel-length SOS MESFETs;

(2)   a study of digital circuit applications for the devices developed in (1);

(3)   a study of linear circuit applications for the devices developed in (1);

(4)   a study of approaches alternative to SOS for device fabrication, e.g., laser-annealed polySi on silicon dioxide;

(5)   a study of the high-field behavior of electron transport in silicon layers on insulating substrates.

During 1979 three different processing methods were developed for short-channel SOS MESFETs: a simple process utilizing SOS material uniformly doped to the desired channel doping during growth; a recessed-channel process developed for power MESFETs, utilizing an ion implant to produce a desired channel profile in intrinsic SOS, and diffusion to produce low-resistance source and drain regions; and a combination implanted/diffused planar process. The last process has so far produced the best devices, with reasonable performance predictable to almost 3 GHz with 0.9-μm channels. Improved performance is expected (from the results of two-dimensional device simulation using CUPID) when source–drain spacing is shortened. These technologies are described in more detail in Barnard et al. [13]. An example of a SOS short-channel device is shown in Fig. 11.

The high-performance planar devices developed at NRRFSS are being incorporated in SSI digital circuits: layout of a 24-transistor divide-by-two circuit has been completed. Our SOS MESFET technology is also being incorporated in high-speed linear circuits. An SOS MESFET differential amplifier with 0.8-μm gate length and four micron source–drain spacing has been fabricated (cf. Fig. 12) and performance is being documented.

The properties of hot-electron transport in layers of both SOS and laser-annealed polySi have been measured. While low-field mobility in

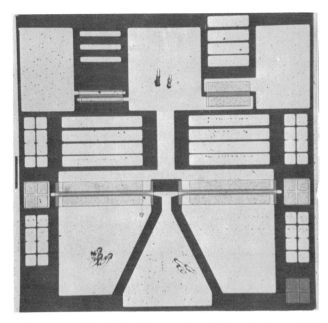

Fig. 11.   Silicon-on-sapphire field-effect transistor. This submicron-gate SOSFET was fabricated by projection lithography, ion milling, and ion-implantation processing.

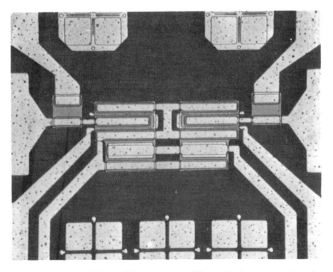

Fig. 12.   SOS MESFET differential amplifier with 1-μm gate length.

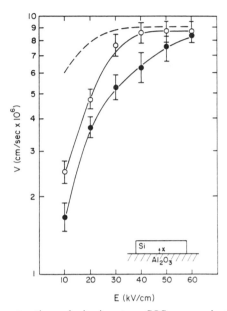

Fig. 13. Electron saturation velocity in n-type SOS versus electric field for $N = 2 \times 10^{17}/cm^3$ at $T = 300$ K. The values shown are for average velocities from the top surface to a depth of 0.4 μm ($\bigcirc$), for the entire layer etched to 0.32 μm ($\bullet$), and for (100) silicon (---).

SOS layers decreases very rapidly as the silicon–sapphire interface is approached, it was determined that saturation velocity remains the same as in bulk Si, i.e., about 10^7 cm/sec. Figure 13 shows an experimental result for a typical SOS layer, thinned down from 0.58 μm and doped during growth to a concentration of $10^{17}/cm^3$. The open circles show the average velocity field relation for electrons in the layer between the top surface and a depth of 0.4 μm; the dark circles show the average velocity in the entire layer when etched down to 0.32 μm thick. Similar data for laser-annealed polySi on oxide indicate that the surface mobility is somewhat lower than that in a layer of SOS equally thick, but the mobility does not greatly vary with depth and the average velocity of electrons in laser-annealed poly, averaged across several crystallite boundaries, can be much higher than that in an SOS film of equal thickness [14]. These results show that both SOS and laser-annealed polySi on oxide have substantial potential for high-speed submicron devices.

2. Optical Devices

A number of optical devices that utilize small lateral dimensions or advanced materials technology based on molecular-beam epitaxy have been investigated by Professor Ballantyne and his co-workers in the School of

Electrical Engineering. The broad objectives of his program are to develop lithographic and materials techniques suitable for fabrication of integrated optical devices and subsystems and to apply this technology to the construction of some representative integrated optical components. Methods of optical holography and optical and e-beam lithography are being developed for applications in optical devices. Appropriate processing techniques for transferring these patterns into substrate materials are also being developed.

One task utilizes optical holography to fabricate gratings for various applications. Gratings with periods from several microns down to 3000 Å are routinely made on either transparent or reflecting substrates and can be used for ion milling or lift-off processes [15]. Work is active on fabricating gratings with periods down to 1000 Å by utilizing glass prisms and liquid tanks to shorten effectively the exposing wavelength. Gratings with periods of 1500 Å have been fabricated but are not yet uniform. Application of these gratings to the fabrication of a polarizer for use in the ultraviolet is being pursued. Optical polarizers with 3000-Å periods have been fabricated and show polarization ratios of 0.4:1 at 6328 Å and 0.8:1 at 4131 Å. Current efforts are directed toward reducing the period to 1500 Å so that substantial polarization for light of wavelength 2000 Å will be achieved. It should be noted that a simple wire grid will not serve as an efficient polarizer; instead the structure should more nearly resemble a stack of short, parallel-plate waveguides.

A second task in this project has involved constructing an optical waveguide integrated with an OPFET detector [16], for which the structure shown in Figs. 14 and 15 was fabricated. All lithographic steps in the fabrication of this device were done on the Kasper 10× optical aligner. The

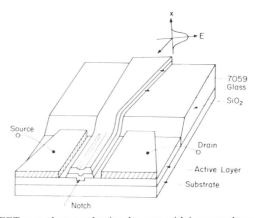

Fig. 14. An OPFET-type photoconductive detector with integrated waveguide structure.

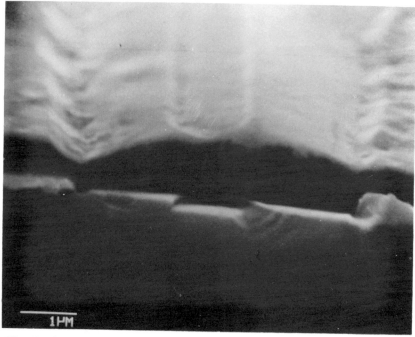

Fig. 15.   SEM photography of cross section of the detector in Fig. 14. The notch (center) is about 1.2 μm wide.

minimum lateral dimension in this device is about 1 μm, but the requirement for smoothness on the walls of the two-dimensional waveguide is more stringent, as are tolerances on all vertical dimensions. The waveguide was formed in sputter-deposited 7059 glass, and its coupling to the underlying GaAs was controlled by the thickness of a sputter-deposited SiO_2 isolation layer. All pertinent device concepts were demonstrated with the units constructed, and the results continue to show that the OPFET is a very promising detector for applications in high-speed optical communications. The device in Figs. 14 and 15 had response times on the order of 100 psec, had internal gain of about 8, and operated at an applied voltage of 3V. This device was also significant in that the gate electrode was eliminated, allowing better optical coupling to the active region and unequivocably establishing the operating mechanism as photoconductivity.

High-speed photodetectors in the configuration of Fig. 14 (but without the waveguide) have also been fabricated in GaInAs grown by MBE on semi-insulating InP substrates [17]. These are attractive because they

respond to wavelengths out to 1.6 μm and GaInAs has a high peak velocity and the largest electron mobility of the GaInAsP system. The GaInAs
detectors had 200-psec speed (FWHM) with internal gain of 10 at an
operating voltage of 2.5 V.

A different structure, which illustrated the application of advanced
MBE materials growth techniques to the fabrication of an optical detector, was the construction of a large-area, transit-time photoconductive
detector using the OPFET operation principle. This device is shown in
Figs. 16a and 16b [18]. An n^+, n^-, n^- graded to n^+ structure is used. The
detector shown in Fig. 16 utilizes a contact window of 1000 Å of super-
doped n^+ GaAs; the active layer is 2 μm of lightly doped (2×10^{15} cm^{-3})
n-type GaAs, and the substrate is n^+ GaAs. The grading of the n^- to n^+
layer between the active layer and substrate is done utilizing atomic plane
MBE doping techniques [19]. This 1.5-μm-thick region actually consists
of 100 separate layers. The device shown in Fig. 16 operated at voltages
less than 5 V with response time (FWHM) about 100 psec and photoconductive gain of 8 [18]. The internal gain is important when the noise of the
external electronics dominates, and the device has the advantage of requiring only low voltage.

Additional projects on integrated optical circuits and devices are being
carried out by Professor C. L. Tang in the School of Electrical Engineering. Two projects underway at the time of this writing were the fabrication
of thin-film Nd^{3+}-glass lasers using resonant grating reflectors and the fabrication of ion-implanted optical waveguides and circuits in LiNbO$_3$.
Tang previously demonstrated [20] a stimulated emission gain of over 1
cm^{-1} at 1.06 μm in Nd^{3+}-doped, thin-film, glass waveguides. To achieve
laser action making use of this gain medium, it is necessary to fabricate
two etched grating reflectors with a period of 0.35 μm separated by 1 cm
or more on the glass waveguide. Substantial work has been accomplished
to prepare Nd^{3+}-doped-glass films of reproducible quality suitable for fabricating this type of laser. A problem of cracking in the glass guides due
to different coefficients of thermal expansion of the Nd-glass film and the
Pyrex substrates used previously has been overcome by utilizing SiO$_2$ on
Si as the substrate. A theoretical study on design constraints, such as film
loss and thickness uniformity, and population inversion necessary to construct the laser has been completed and laser action in the structure
should be achieved shortly. The anticipated benefit of Ti ion-implanted
waveguides in LiNbO$_3$ is a lower process temperature, which will avoid
the problems associated with surface out-diffusion during the high temperatures necessary to fabricate diffused waveguides. Work is also underway to fabricate two-dimensional, diffused waveguides on LiNbO$_3$
and various optical switches.

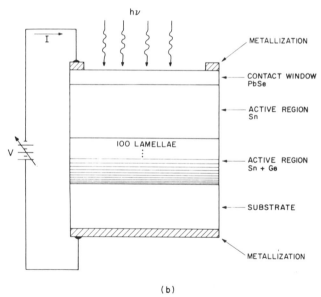

(b)

Fig. 16. (a) Photomicrograph of an experimental transit-time photoconductive detector. The contact pad is 75 μm × 75 μm and is isolated from the n⁺ substrate by 1 μm of sputter-deposited SiO_2. (b) Structure for the large-area, transit-time photoconductive detector. This represents a cross section of the circular region in (a). Dopants for the various layers are listed.

3. Josephson Junctions

Professor Buhrman's group has applied the high-resolution fabrication techniques described previously to the construction of Josephson junction devices. The major result during 1980 was the development of a versatile technique for the production of rugged submicron Josephson tunnel junctions of $Nb-Nb_2O_5-PbBi$. By forming junctions on the fabricated edge of a Nb film, submicron area junctions (0.1 μm^2) can be formed in a manner compatible with standard thin-film processing (see Fig. 17). By developing a new ion-beam oxidation process, Josephson junctions of unsurpassed quality have been produced. Because of their extremely small size, such junctions are highly suitable for a number of device applications, including millimeter wave mixing. These applications are now being actively pursued.

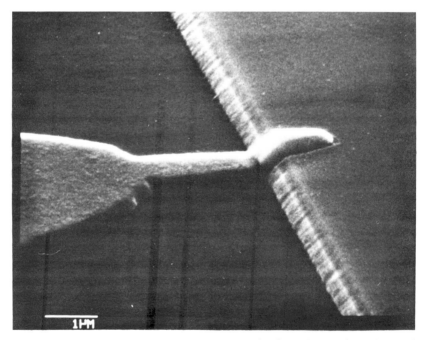

Fig. 17. A $Nb-NbO_x-PbBi$ Josephson tunnel junction formed on the faceted edge of a Nb film. Contact between the PbBi top electrode and the Nb base electrode is eliminated everywhere but at the Nb edge through use of an Al_2O_3 insulating layer on top of the Nb electrode.

4. Electrode Arrays

There are several projects underway in the facility that utilize submicron lithography to construct electrode arrays for making measurements on various physical systems. One such project under the supervision of Professor J. Osteryoung at the State University of New York, Buffalo, involves electrode arrays for studying size effects in electrochemical reactions. Another is a project by Professors W. Heetderks and M. Kim in the School of Electrical Engineering at Cornell to construct LSI electrode arrays for recording from populations of neurons. A third program for fabricating small electrodes is the work of Professor Ruoff on interdigitated electrodes for high-pressure measurements.

Such arrays as in the project by Professors Heetderks and Kim are necessary because the study of the nervous system even in simple invertebrates is ultimately the study of a highly complex system. A small ganglia or cluster of neurons can easily involve hundreds of cells, and these cells are interacting as a total system to produce different motor responses to different input stimuli. The first step in studying the operation of a simple nervous system is to study the individual cells that make up the nervous system. The appropriate second step in the analysis should involve the study of the response of the entire population of cells or the entire system.

Considerable work has been directed at observing the response of many cells by using signal-processing methods to distinguish different cells on one electrode on the basis of the shape of the action potential. Using such techniques, one electrode can be used to record from about five cells. In order to record from greater numbers of cells, multiple-electrode arrays are required.

Multiple-electrode arrays might be based on a variety of artifacts associated with action potentials. Associated with each action potential in a cell is a sudden change in the potential across the cell membrane of about 100 mV. Also, the cell size changes very slightly; some cell dyes change their absorbance very slightly. Two approaches to observing the action potential are to measure extracellular currents associated with the potential change or to measure changes in absorbance (or fluoresence) of dyes that are applied to cell membranes. Both of these approaches are well suited to multiple-recording arrays. For the optical measurement, the multiple-recording sites correspond to the multiple points on the surface of an imaging device. The major problem with the optical technique is finding a suitably sensitive dye. When extracellular currents are measured, sensitivity is not a major problem. The problem is to fabricate an array of electrodes for recording.

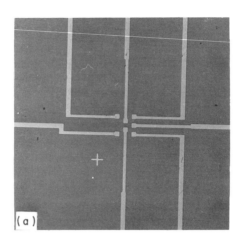

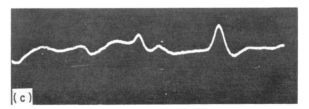

Fig. 18.

The microelectrode has been used for 40 years to study the response of single cells in the nervous system and is still widely used. An electrode array capable of recording from nervous systems as opposed to single cells should have similar utility. The goal of this work is to produce an extracellular electrode recording array that has the following elements:

(1)   an array of recording electrodes,
(2)   an array of preamplifiers to buffer the signals from the recording electrodes, and
(3)   a multiplexer system that permits the signals from many electrodes to be multiplexed onto a single line.

The present goal is to design and fabricate an array of 10×10 electrodes with $100\text{-}\mu\text{m}$ spacing, each electrode having a $10\text{-}\mu\text{m} \times 10\text{-}\mu\text{m}$ recording area. Within the $100\text{-}\mu\text{m} \times 100\text{-}\mu\text{m}$ space associated with each electrode, a preamplifier and a multiplexer will be integrated. Later the array may utilize submicron feature sizes.

At this writing, work is progressing in three areas. A seven-electrode array with no active on-chip electronics has been fabricated. These electrode chips are being used to evaluate chip-insulation techniques and electrode-recording techniques. Passivation techniques being tested include two sputtered glasses, one spin-on glass, and a baked photoresist. The electrode is shown in Fig. 18a and the bonded, insulated electrode shown in Fig. 18b. Figure 18c shows action potentials recorded from a cockroach nerve using the electrode array. The nerve was raised from the abdomen and set against the electrode for recording.

The second area of work involves fabricating a test chip containing JFETs of various sizes. The gates of some of the JFETs on this chip are connected to electrodes so this chip can also be used to evaluate the characteristics of the amplifier in a biological environment. Wafers are currently being processed.

The third area of work being done involves designing an amplifier and multiplexer. Although it is difficult to design a circuit before the device characterization previously mentioned is done, JFET characteristics has been calculated based on the geometry (assuming reasonable process control). The design requires that each electrode amplifier and multiplexer unit plus the bus lines fit in a $100\text{-}\mu\text{m} \times 100\text{-}\mu\text{m}$ area. The amplifier–multiplexer has been designed, simulated on SPICE, and breadboarded. The circuit appears to have all the characteristics required for the amplifier multiplexer.

Fig. 18. Electrode arrays for neural studies: (a) the seven-electrode array for testing chip insulation and recording techniques; (b) the same array mounted for recording; (c) action potentials recorded with (b) from a cockroach nerve.

The program by Ruoff is an example of how small structures fabricated in the facility can be used to facilitate basic measurements in other fields. This research is in the general area of phase transitions in materials at very high pressures. Pressures on the order of 1 Mbar or greater are most conveniently attained in very small volumes. Hence, measuring instruments are needed that will fit within the attainable pressure volume. Ruoff and his co-workers have used the approach of creating high pressures between a flat diamond anvil and a spherical diamond indentor with very small tip radius. For pressures over a megabar, interdigitated electrodes with a period less than 1 μm are required [21]. This technique was applied by Ruoff and Nelson [22] in experiments that produced metallic xenon. The technique not only requires less force than traditional methods because of the small pressure volume but has the advantage that the interdigitated electrodes do not short together (as would

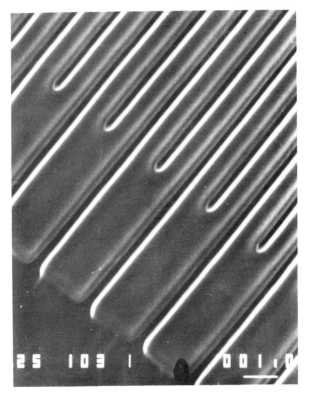

Fig. 19.   Interdigitated electrode pattern in PMMA for fabricating electrodes on diamond for high-pressure experiments. Linewidth is about 0.2 μm on 1-μm centers.

flat plates separated by the sample) if the indentor punches through the sample. The interdigitated electrodes, unlike flat-plate electrodes, are not particularly sensitive to the positioning of the indentor. An example of an interdigitated electrode pattern produced for Ruoff's experiments using e-beam lithography in the facility is shown in Fig. 19.

C. Materials Research and Fundamental Studies

The performance of semiconductor devices is intimately related to the properties of the material from which they are constructed. As the dimensions of the device are made smaller and smaller, the materials begin to behave differently when the device dimension approaches characteristic scattering lengths, and one can begin to think in terms of new material structures applicable in devices with very small dimensions. Programs in the Submicron Facility treat aspects of this problem.

1. III–V Compounds

Professor L. Eastman and Dr. C. E. C. Wood in the School of Electrical Engineering are pursuing research on molecular-beam epitaxy of III–V compounds and its application to various devices. Studies of tin dopant incorporation have led to GaAs FET active layers, with exponentially decreasing, free donor profiles being produced by depositing all of the required dopant on the substrate surface prior to growth [23]. Devices with such active layer profiles have very linear transfer characteristics, which agree with theoretical predictions (see Fig. 20).

Deposition of dopants such as germanium while growth is periodically suspended is being used to generate complex free donor profiles in a digital fashion known as *lamellar doping.* FET active layers have also been grown in which sufficient free carriers are produced from one dopant lamella at the substrate active layer interface [24]. Electron mobilities at 77 K in these layers are more than four time those observed in conventionally doped layers. Device noise should thus be significantly reduced. This process of lamellar doping was applied to synthesize the particular linear doping profile required for optimizing performance of the optical detector described in the preceding section. MBE growth was also used to produce the InGaAs material for the long wavelength, photoconductive detector previously described, and similar material has been used to produce microwave MESFET transistors that have higher current gain and frequency than the best produced previously at Cornell in GaAs. Current work is underway to perfect new heterojunctions of InGaAs and InAlAs on InP substrates for use in microwave and optical devices.

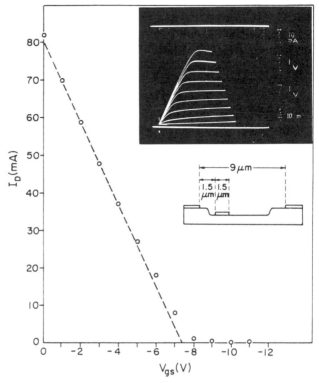

Fig. 20. Molecular-beam epitaxy is used to produce FET active layers of GaAs with exponential doping profiles. These are measured linear-transfer characteristics of a GaAs field-effect transistor produced from the MBE-grown active layers.

In addition to materials growth, an extensive program is underway to characterize the impurity levels and concentrations in the grown layers using photoluminescence, deep-level transient spectroscopy (DLTS), Auger profiling, secondary-ion mass spectroscopy (SIMS) measurements, and Rutherford backscattering.

A program is underway in the same group to do molecular-beam epitaxy of GaAs in selected submicron areas defined by previous lithographic processes. Materials for oxide masking of the GaAs during MBE growth are being developed and test patterns have been designed to allow growth of layers in selected areas geometrically useful for Hall, C-V, DLTS, optical, and structural assessment and evaluation of both the epitaxial single-crystal areas and the polycrystalline nonepitaxial GaAs regions of layers and the single/poly interfaces. These studies will be done as a function of the selected area dimension and layer thickness.

2. Interface Studies

Related work in materials and interface characterization is being carried out by Dr. L. Rathbun and Professor T. Rhodin from the School of Applied and Engineering Physics. The Submicron Facility is equipped with a Physical Electronics 590 Scanning Auger Microprobe (SAM). This automated instrument allows routine analytical characterization with a spatial resolution of 200 nm. The microprobe has been retrofitted with a sample introduction/reaction chamber. This not only allows rapid turn-around of routine samples but opens the possibility of performing in situ surface reactions, sample treatment, and characterization. Research on the Auger microprobe encompasses two rather distinct areas: collaborative research in support of studies initiated by others and core research on surface and interface phenomena initiated by Rhodin and Rathbun.

The core research concentrates on UHV studies of atomic processes in microfabrication. The controlled generation of microstructure is ultimately determined by processes on an atomic scale. These include, among others, the adsorption, desorption, nucleation, and diffusion of atoms and molecules on solid surfaces. Often, little is known about how these fundamental events relate to the actual technological process. The core research utilizes scanning Auger and other surface techniques to study these events on semiconductors. Some of the important areas being addressed are mechanisms in dry chemical etching, nucleation and growth of two-dimensional metal overlayers, and surface and interface diffusion in metal–semiconductor systems.

The collaborative research involves application of Auger characterization techniques to a wide variety of problems in submicron technology. Part of this work is devoted to the development of techniques and standards that will allow accurate quantitative microchemical characterization of compound semiconductors using Auger electron spectroscopy. The work is aimed at allowing one to understand and account for two basic effects. The first is variation due to selective sputtering, which results in a surface composition that differs from the bulk. Auger, which measures only the surface composition, is particularly sensitive to this effect. The second effect, electron backscattering, means that Auger electrons are excited at the surface, not only by incident electrons, but also by electrons backscattered deep within the solid. The number of Auger electrons is thus a complex function of the composition of the solid. Accurate quantitative analysis requires that the separate contributions of these two effects be clearly understood. At the time of this writing, calibration data had been developed for the determination of indium concentration in various compound semiconductors. Standard samples of $Ga_xIn_{1-x}As$

were produced by molecular-beam epitaxy and a study had been initiated
to separate the results of selective sputtering from those of backscattering. Bulk binary compounds being studied included GaAs, InAs, and
GaP.

Microchemical defects in GaInAs and GaAs junctions produced by
molecular-beam epitaxy were identified by Auger microanalysis and secondary electron emission patterns [25]. Nonuniformity in the films was
shown to be associated with "spitting" effects in the vapor source. Com-
positional inhomogeneities led to geometrical defects in the junctions and
should be avoided during formation of the layers, hence this information
led to changes in the MBE growth procedures used in the facility. In another study [26], the composition and degradation of GaInAs–AlInAs
heterostructures were studied as a function of growth parameters. The
microanalytical capability of the Auger microprobe was critical to understanding the growth process in this work.

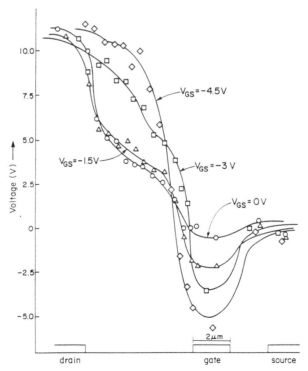

Fig. 21.   Scanning Auger voltage profile across GaAs MESFET at a 12-volt drain–source
bias as a function of gate bias.

The use of the Auger spectrometer in quite a different mode was illustrated by measuring voltage drop profiles across metal–semiconductor FETs developed for microwave use [27]. Using the scanning Auger microprobe, the variation in potential across an operating device can be determined to ± 0.2 eV with a resolution of 200 nm. This is done by monitoring shifts in a carbon Auger peak as a function of applied voltage. The objective of the work was to increase the working breakdown voltage by identifying the regions of high, local, electric field within the FETs. These profiles were very effectively determined in the Auger microprobe and have provided data leading to FET devices with enhanced high-voltage characteristics. An example of these measurements is shown in Fig. 21.

3. Analysis on the Nanometer Scale

Electron energy loss is a very promising technique for chemical analysis of extremely small volumes. When high-energy electrons such as those in the probe of a STEM pass through a solid sample, they lose energy to the atoms comprising this solid in amounts characteristic of the atom from which they were scattered. These energy losses can be measured and used to deduce the concentrations of different kinds of atoms scattering the electrons. The electrons also lose energy to the characteristic electronic excitations of the solid and, hence, energy-loss spectroscopy in an instrument such as the STEM can provide both chemical and electronic characterization in the 1–10-nm region. Since silicon is a prime material for device manufacture, development of quantitative, STEM, thin-film studies is important in the high spatial resolution regime. Professor J. Silcox in the School of Applied and Engineering Physics is pursuing work in this area and has initiated work on the preparation and characterization of thin films of silicon in the 200–600-Å thickness regime.

The films were prepared by an extension of the technique of chemical etching of boron-doped silicon first used by Huang and Van Duzer [28]. This technique used a boron concentration of $7 \times 10^{19}/cm^3$ to stop the etch, thereby limiting the thickness of the films. Two techniques were used for introducing the boron into the silicon: diffusion from liquid B_2O_3 on the surface of the wafer and ion implantation into an oxide coating on the surface of the wafer, followed by diffusion. When liquid B_2O_3 was used as a diffusion source, the films consisted of plateaus, each plateau uniform in thickness but with the thickness of the film varying between 600 Å in the thinnest regions and 1400 Å in the thickest of the plateaus. Holes were also present in the film, and bend contours assumed highly anfractuous shapes, suggesting the presence of strain in the film.

The use of ion implantation enabled the controlled fabrication of films that were about 250 to 300 Å in mean thickness. However, even these films, when examined on a scale of 200 Å, exhibited fine-grained nonuniformities in thickness, which could presumably be attributed to the fluctuations in the boron concentration required to stop the etch. A higher concentration of dislocation loops was observed in the implanted specimen than in the specimen doped from the B_2O_3 source, which may be due to either one or any combination of the following factors: implantation damage, lattice strain due to the alteration of the silicon lattice constant by the boron, and strain at the oxide–silicon interface during the diffusion, due to the significant difference in the thermal expansion coefficients of silica and silicon. Although an attempt was made to minimize the effect of the factors contributing to the damage in the film by implanting the boron into the oxide near the oxide–silicon interface and subsequently diffusing the boron into the silicon, it was found that the damage and strain in the film was not entirely eliminated. In an effort to produce damage-free films, current work is focused on the development of a thinning technique based on anodic etching [29].

Characterization of the silicon films proceeded using the STEM. Figure 22 represents a bright field image of the ion-implanted specimen, illustrating damage in the form of small dislocation loops. Figure 23 is an annular dark field image of the sample doped by diffusion from a liquid B_2O_3 source. The level of brightness in this image corresponds to the mass thickness of the sample [30] and thus is a measure of the specimen

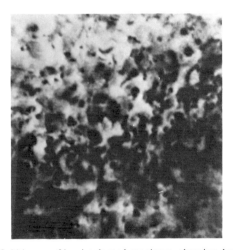

Fig. 22.   Bright field image of ion-implanted specimen, showing damage. Horizontal field of view is 1.36 μm.

Fig. 23. STEM annular, dark field image of specimen doped by diffusion from a liquid B_2O_3 source. Horizontal field of view is 13.8 μm.

thickness (about 1400 Å in the brightness area and 600 Å elsewhere). The crystallographic anisotropy of the etching process is revealed by the shapes of the etch figures. Note also that the holes project in (110) directions in the plane of the sample.

Silcox, Lee, and Fejes have also explored the use of convergent-beam electron diffraction as a tool with which to measure specimen thickness from microareas of the order of 20-Å diameter, following the work of Goodman and Lehmpfuhl [31]. Figure 24 is a high-magnification image of

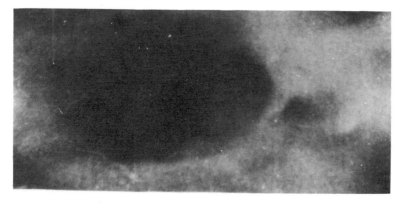

Fig. 24. High-resolution, bright field image of ion-implanted specimen in Fig. 22. Horizontal field of view is 0.196 μm.

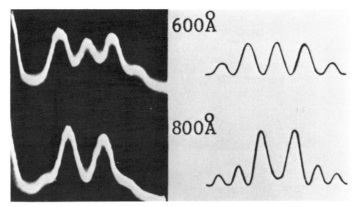

Fig. 25. Comparison of experimental with theoretical scans of intensity across convergent beam diffraction patterns taken from two areas of Fig. 24.

the film shown in Fig. 22. In order to verify that the variations in intensity across the field of view were indeed due to variations in thickness of the specimen, intensity scans across convergent-beam diffraction patterns were obtained from various points on the specimen, and these were compared with theoretical curves produced by an eleven-beam dynamical Bloch wave calculation. Given accurate values of scattering factors available from Lehmpfuhl's work, they expect eventually to be able to measure the thickness to within one unit cell with this technique. Further, it can be used for calibration of the annular, dark field signal which may be more convenient to use in practice. In Fig. 25 is shown a comparison between the calculated and experimental scans of the diffraction pattern for thicknesses of 600 and 800 Å. The variations in intensity in Fig. 24 correspond to variations in thickness between 250 Å in the light areas to 800 Å in the darkest areas.

Another example of the use of the STEM for analytical work was Professor Isaacson's success in obtaining electron diffraction patterns from a single macromolecule of ferritin. The molecule is 120 Å in diameter with a core about 60 Å in diameter that can be filled with up to 5000 iron atoms in a crystalline, mineral form. Electron diffraction patterns were successfully obtained from the core and are being used to determine the crystal structure of the iron in the core.

In a similar vein, the STEM was used by Dr. P. E. Batson, a Visiting Research Fellow from IBM Thomas J. Watson Research Center, to investigate the plasmons in small aluminum spheres (5–50-nm diameter). The width of the bulk plasmon was found to change with sphere size, as shown in Fig. 26. This width variation results from a decreased electron velocity

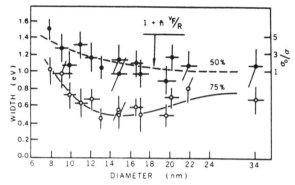

Fig. 26. Width of the volume plasmons in small Al spheres as a function of sphere diameter at 50 and 75% of full height. The dashed line indicates the expected behavior for electron–wall collision damping. The derived bulk conductivity, relative to the large sphere conductivity, is indicated on the right.

relaxation time brought on by the proximity of the sphere walls. Therefore, this relaxation time may be measured and, through simple Drude ideas, related to the dc conductivity. This result is described in Batson [32] and is also shown in Fig. 26. Batson has, therefore, shown that it may be possible to make noncontacting electrical tests such as conductivity measurements on very small structures.

4. Other Studies

Several additional projects on various fundamental problems were just getting underway as of this writing. They included studies of quantum size effects in samples of metals and insulators at very low temperatures, studies of the mechanism of heat transfer to very small objects, a study of weak-link and Josephson behavior in superfluid ^{3}He, and an examination of equilibrium and transport properties of very small metal particles, including possible cooperative effects in large aggregates of such particles.

D. Circuit Design and Layout

Most work in this area was initiated at Cornell subsequent to the coming of the national facility; hence, much of it is still in an early stage.

1. Integrated Circuits

Some of the device work described in Section II.B has now progressed to the point where devices with submicron dimensions are beginning to be incorporated into small-scale integrated circuits. An example is the gallium arsenide integrated circuit program in Professor Eastman's group. The general objectives are to demonstrate small, high-performance modules for later incorporation into medium- and large-scale integrated circuits. Configurations that have been experimentally fabricated and investigated are discrete FETs, inverters, ring oscillators, NAND and NOR gates, and microwave amplifiers. The objectives are somewhat parallel to those of Professor Frey's program on silicon-on-sapphire devices and circuits described in a preceding section. Current work in the GaAs IC program utilizes projection photolithography with minimum gate lengths of 0.75 μm. Within three years it is expected that direct e-beam writing will allow shrinking linear dimensions to 0.2-μm gate lengths.

This program is also using an innovative materials technology for circuit fabrication. Circuits are fabricated on layers grown by molecular-beam epitaxy over single-crystal substrates patterned with a growth resist. A high-temperature native oxide is currently used as the growth resist and seems to be very satisfactory. In areas where a device is to be fabricated an opening is made in the growth resist so that the subsequent MBE layer will be single crystal. In areas that are to be passive or used for circuit interconnects, the resist is left on the substrate so that the grown MBE layer is polycrystalline and, hence, has high resistivity. The actual resistivities in the polycrystalline areas exceed 10^5 Ω cm with a breakdown field of about 10^5 V/cm (for $n = 1-5 \times 10^{17}$ cm^3). Resistivity in the single-crystal regions is, of course, controlled by the doping in the usual manner. Because the difference in height of the single and polycrystalline areas is only 5% of layer thickness, this approach results in essentially a planar process technology, with the single-crystal islands for active devices being separated by an insulating polycrystalline sea. Schottky diode/FET NAND and NOR logic elements have been fabricated and operated using this approach [33]. A schematic diagram of the layer cross section is shown in Fig. 27, and Fig. 28 shows an SEM micrograph of a selective-area MBE wafer for a nine-stage, depletion-mode, ring oscillator. Current performance goals for this project include 50-psec/gate switching speeds and 40-GHz microwave amplifiers. The longer-range targets that appear to be achievable are 10-psec/gate switching times for a fanout of 1 and less than 1-fJ power delay products. Enhancement mode logic will be used to achieve these low-power delay products.

Another example of a discrete device program growing into integrated circuits is the integration of the fast transit-time photoconductors described in a previous section with preamplifiers. Because of the high speed and low noise of GaAs transit-time photoconductors, it is impossible to take full advantage of them without very sophisticated preampli-

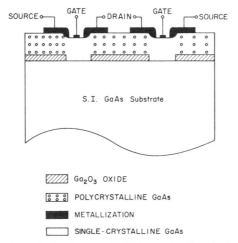

SOURCE○— GATE ○ —○DRAIN○ GATE ○ ○SOURCE

S.I. GaAs Substrate

▨▨▨ Ga₂O₃ OXIDE

POLYCRYSTALLINE GaAs

METALLIZATION

SINGLE-CRYSTALLINE GaAs

Fig. 27. Cross section of selective-area MBE layer showing single and polycrystalline areas.

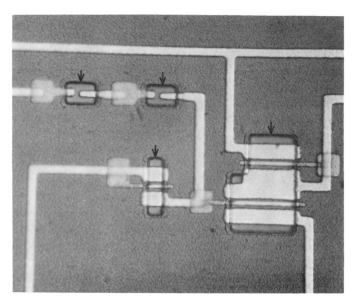

Fig. 28. Photomicrograph of one stage of a nine-stage depletion mode ring oscillator. The single-crystalline epitaxy areas for the active devices of the ring oscillator are indicated by arrows. Gate length is nominally 1 μm.

fiers. The noise levels and speed of the detectors are comparable to the best X-band, GaAs, small-signal MESFETs. A new program is, therefore, underway on the design and construction of monolithically integrated transit-time photoconductive detectors with GaAs FET preamplifiers. Eventually such an integrated optical detector and amplifier combination could become a building block for incorporation into larger monolithic systems employing the logic elements previously described.

2. Computer Systems

Larger-scale systems are also being studied in computer systems research within the Submicron Facility. Part of this effort relates to computer-aided design. In addition to maintenance and enhancements of the Applicon SVLSI CAD system for e-beam pattern generation, other projects to enhance the system are underway. Working under RSX-11 and UNIX, Professor Hammerstrom in the School of Electrical Engineering and co-workers are designing software packages and a comprehensive data base for high-level, hierarchical, digital system design. The data base will efficiently integrate both the functional and the physical information of a digital system. The combined functional/physical data base will be very useful for electron-beam testing at the system level since it provides the crucial link between functional information about the circuits and their physical construction and location. The program on system-level electron-beam testing is a new one being initiated by Professors Wolf and Hammerstrom.

The system research also addresses problems related to interconnection analysis. Traditional analysis techniques and design methodologies are no longer relevant in the area of VLSI computer engineering. Consequently, new design techniques, as well as cost measures, are needed. It has been demonstrated that a hierarchical interconnection structure is more efficient (it minimizes the area time product) for integrated circuit fabrication than other types of commonly used structures. The goal of this research activity is to develop new means of interconnection analysis. A hierarchical or "clustered" interconnection structure is assumed and Hammerstrom's group is applying information-theoretic techniques to this structure, since information flow is what is to be measured. A goal is to develop design methodologies that are based on these techniques and optimize operand placement and interconnect structure in a two-dimensional silicon system.

A further program on distributed computing is underway. The VLSI systems of the future will be, due to pin constraints, created out of loosely coupled, self-contained computational units at the chip or board level.

These units will, in turn, share expensive secondary memory and input/output devices. To be able to use such systems effectively, an efficient method for leveling the load on the network (moving computational tasks to idle processors) is needed. Ideally, such load-leveling should proceed under decentralized control. To load-level optimally is very difficult and cannot be guaranteed under decentralized control. However, a heuristic procedure has been developed [34]. This technique, though assuming a static environment, is decentralized (implemented independently at each processor) and dead-lock-free. The procedure often finds the optimal allocation and is always close to optimal (usually within 5 to 10%). The next step is to expand the procedure to a dynamic environment where task behavior changes with time and tasks are dynamically created and terminated. In this environment is is important to still effect load-leveling, but now the possibility of network instability must be considered. Consequently, one goal is to derive the necessary conditions to guarantee stability within the network. Also, task performance estimation procedures are being developed. An additional goal of the computer systems research activity is an experimental implementation of the above techniques on the Submicron Facility's computer network to verify their effectiveness and to enhance the network's computational capability.

III. RESOURCES

A. Instrumentation and Facilities

A major objective of the facility is to provide an equipment resource for submicron structures fabrication and research. At this writing, the value of the equipment resource generated by the NSF grant and the industrial affiliates program (called PROSUS) exceeds $4.0 million. This equipment investment covers most of the required instrumentation for microstructures research and development. It provides advanced capabilities in electron, ion, and photon lithography; thin-film materials preparation, patterning, and evaluation; advanced device and circuit fabrication and test, as well as exploratory nanostructure fabrication and research. All of the listed equipment is installed and operating and, at this writing, most of the instruments are in temporary laboratories on the fourth floor of Phillips Hall. Table III gives a list of the major equipment included in the facility at this time. Figure 29 shows a photograph of the molecular-beam epitaxy system in the existing clean room on the fourth floor of Phillips Hall which contains also the x-ray system and the ion-beam milling

TABLE III

Equipment Resources at the NRRFSS

Equipment	Uses
Pattern generator	Computer-Aided Design System: Applicon 860 Super VLSI. Interactive design and drawing digitizing capability. Resolution of 1 part in 10^9.
	Electron-Beam Pattern Generator: Cambridge EBMF 2–150. Resolution of 0.2 μm or better in 1-mm $\times$ 1-mm field. Adjacent fields stitched together with 0.3-μm accuracy. Pattern size up to 4-in. dia. Multilayer registration to 0.3 μm.
	Experimental STEM Pattern Generator: Vacuum Generators HB5. Resolution about 100 Å or better over small fields.
	Optical Holography: Generation of gratings with 0.2 μm periods.
	Exploratory Ion-Beam Lithography (1981).
Pattern replication	Optical Step-and-Repeat: D. W. Mann. X-Ray lithography systems: Low-intensity sources for pattern resolution 100 Å or better. Synchrotron radiation (available in 1981) for fast exposure, with resolution of 0.1 to 0.2 μm.
	Optical Proximity and Contact Aligners: Cobilt CA 400 A. Resolution of 1 μm over 3-in.-dia. wafer.
	Optical Projection Aligner: Kasper 2510. Resolution of 0.7 to 1 μm over $\frac{1}{4}$ in. $\times$ $\frac{1}{4}$ in. area. 10 $\times$ reduction. Noncontact printing with 5 $\times$ 5 step-and-repeat matrix.
Materials preparation and thin films	Liquid-Phase Epitaxy Systems: Layers of III–V compounds 0.1–10 μm thick over 2-cm^2 area.
	Molecular-Beam Epitaxy System: Varian 360. Layers of III–V compounds with thickness control down to 100 Å.
	Thermal and Electron-Beam Evaporators: Veeco, CVC, Varian.
	DC and RF Sputtering Systems: Up to 6-in.-diameter targets and substrates; up to three materials sequentially. Sputter etching.
	Ion-Beam Deposition and Milling System: Commonwealth Scientific 4-in.-diameter beam. Up to four materials sequentially. Etching.
	Reactive Plasma Etch System: 8-in.-diameter planar configuration.
Doping	Hot Tube Diffusion System: 2-in.-diameter wafers.
	Ion-Implantation System: Accelerators, Inc. Model 300 R. 300 keV, all elements in periodic table. 3-in.-diameter wafers. 600 keV for doubly ionized species.
Device processing	Thermal Oxidation Tubes (steam, TCE). Wire and die bonders, wet chemical processing stations, automatic diamond scribers, wire and diamond saws.

TABLE III (*continued*)

Equipment	Uses
Device and structure analysis	Scanning Electron Microscope: Cambridge S150. 70-Å resolution. Up to 3-in.-diameter samples.
	Scanning Transmission Electron Microscope: Vacuum Generators HB5. 3-Å resolution. Elemental analysis by energy-loss spectroscopy.
	Scanning Auger Microprobe: Physical Electronics Model 590. Resolution to 0.2-μm lateral, 10-Å depth. Sensitive to 1 part in 10^3. Computer-controlled analysis. Depth profiling.
	Ion Microprobe: Cameca. Resolution of 1-μm lateral, 10-Å depth. Sensitivity up to 1 part in 10^7. Computer-controlled analysis. Depth profiling.
	Surface Profilometer: Dek-Tak. Depth resolution to 100 Å. Lateral scan to 3 cm.
	Scanning Optical Microprobe: Areal resolution less than 1 μm. Wavelengths 1.06–0.33 μm.
	Optical Microscopy: Leitz, Olympus, Bausch and Lomb.
	Rutherford Backscattering System (1981).
Computational resources[a]	APPLICON SVLSI (AGS 860) Computer-Aided Design System.
	DEC PDP-11/34: 94KW; RP06 (176MB) disk drive; floating point processor.
	Special-purpose (32-bit) graphics processor.
	CONRAC Raster Scan CRT (graphics display and controller).
	Large and small digitizer tablets.
	Dual denisty (800/1600 bpi) tape drive.
	DEC PDP-11/60: 64 kW; 2 RK07 (28MB each) disk drives; RL01 (5MB) disk drive; floating point processor; multiplexer (8 line) with terminals.
	DEC PDP-11/34 (primarily for interface to the STEM): 64 kW; 2 RL01 (5MB each) disk drives; floating point processor; multiplexer (8 line) with terminals; DR11-B parallel line interface unit (DMA).
	Multimicroprocessor system: 3 LSI-11 microcomputers; 16 kW each; 3 DLV11 serial line interface units; 4 DRV11 parallel line interface units; RX02 (1MB) floppy disk unit.

[a] Note: All systems run Digital Equipment's RSX11-M (except for the LSI-11 system, which can use either Mini-UNIX or Digital Equipment's RT-11 V.3); all systems are linked into a functional computer network under Digital Equipment's DECnet II. This list does not include several dedicated computers that are built into larger pieces of equipment.

Fig. 29. MBE system installed in 4th floor clean room in Phillips Hall.

system. With the exception of the MBE system and the Cameca second-ary ion mass spectrometer, all of the major equipment in the facility will be moved in 1981 into a new laboratory, which will become the permanent home of the facility. The floor plan of this laboratory is shown in Fig. 30. The new laboratory is being constructed at a cost of about $3.8 million, which is being borne by Cornell University with approximately $1.7 mil-lion already secured by gifts to the construction funds. This new wing of Phillips Hall (shown in Fig. 31) was carefully designed to provide a vibra-tion-, electromagnetic interference-, and dust-free environment to house the fabrication and process facilities. In order to minimize effects of elec-tromagnetic interference on beam writing instruments, the sources of such interference were identified, measured, and located at large dis-tances from the critical areas. Careful attention was given to the vibration environment so that major sources of vibration in support machinery were specially mounted, and sensitive lithographic equipment is located on 0.76-m-thick concrete floor slabs, which rest on a special compaction of earth and are isolated from the main building structure.

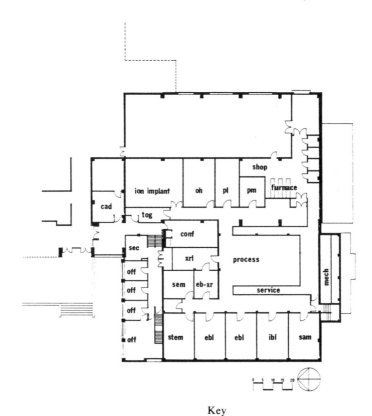

Key

cad	Computer-aided design and computer facility
tog	Dressing room for clean area
oh	Optical holography
pl	Photolithography
pm	Photomask fabrication
xrl	X-ray lithography
sem	Analytical scanning–electron microscope
eb-xr	Resist application and development for e-beam and x-ray lithography
stem	Scanning transmission electron microscope
ebl	Electron-beam lithography
ibl	Ion-beam lithography
sam	Scanning Auger microprobe

Fig. 30. Floor plan of new National Submicron Facility, Cornell University. This facility, scheduled for completion in early 1981, will contain approximately 7,500 ft² of class room and 10,000 ft² of cleaner laboratory and processing space on a single level, and 9,000 ft² of office and support space on two levels.

Fig. 31.   Artist's rendering of new facility's laboratory and offices.

B. Faculty and Staff

Another function of the facility is to provide a resource of expertise in the field of microstructure science and engineering. At Cornell this resource consists not only of a large group of faculty, plus their research groups, which include research associates, graduate students, and support personnel, but also a central staff of the facility proper. A list of the Cornell faculty members who are members of the facility is shown in Table IV, along with their research interests. At the time of this writing, there were ten visitors and research associates and eighteen graduate students associated with these faculty members in their facility-related research. Besides the director and office staff, the facility staff consists of four senior research associates (two shared with other grants), with part-time participation by others in the preceding group, three engineers, and three technicians. The central staff will be increased to provide better research and operational assistance on major equipment installation within the facility as additional funding is secured.

C. Information Resources

Another main function for the National Submicron Facility is to be an information resource to the scientific community. This objective has been met in a number of ways including talks given by facility staff and Cornell

TABLE IV

Cornell Members of the Submicron Facility and Their Research Interests

Members	Interests
Applied and Engineering Physics	
Boris W. Batterman	X-Ray and Neutron Diffraction, Synchrotron Radiation
Robert A. Buhrman	X-Ray and Electron-Beam Lithography, Ion-Beam Processing, Superconducting Devices
Michael Isaacson	Electron Optics, Scanning Transmission Electron Microscopy, Nanometer Lithography
Thor N. Rhodin	Physics and Chemistry of Semiconductor Interfaces
Benjamin M. Siegel	Electron and Ion Sources, Optics and Lithography
John Silcox	Microscopy and Microanalysis with Electron Probes, Electronic Structure
Chemical Engineering	
Ferdinand Rodriguez	Polymer Resists
Chemistry	
George H. Morrison	Microanalysis, Ion Microscopy, Ion Microprobe Studies
Computer Science	
Fred B. Schneider	Design of VLSI Signal Processing Structures and Other Dedicated Systems
Electrical Engineering	
Joseph M. Ballantyne	Electron Lithography, Semiconductor Growth, and Devices for Integrated Optics
G. Conrad Dalman	Microwave Devices and Circuits
Lester F. Eastman	Semiconductor Growth, Molecular-Beam Epitaxy, Microwave Semiconductor Devices
Jeffrey Frey	Microwave Transistors and Integrated Circuits, Physics and Technology for VLSI
Daniel W. Hammerstrom	Distributed Computer Systems, Load-Leveling Computer System Architecture, Computer-Aided Design and Test System for VLSI
Charles A. Lee	Ion Implantation and Lithography, Molecular-Beam Epitaxy, Material and Device Physics
Chung L. Tang	Holography, Devices for Integrated Optics, Nonlinear Optics
Edward D. Wolf	Electron-Beam Lithography, Reactive Plasma Processing, Low-Voltage Contrast for VLSI Testing
Human Ecology	
Sharon K. Obendorf	Analysis of Polymer Films
Materials Science and Engineering	
Dieter Ast	Amorphous Metals
Edward J. Kramer	Strain Displacements in Solids, Superconductor Flux Pinning

(table continues)

TABLE IV *(continued)*

Members	Interests
James W. Mayer	Metal Silicides Interfaces, Ion Implantation, Rutherford Scattering
Arthur L. Ruoff	Ultrapressure Research, X-Ray Lithography, Electron Lithography, Mechanical Properties
	Physics
James A. Krumhansl	Small-Systems Physics, Quantum Effects
Robert C. Richardson	Millidegree Kelvin Research
John D. Reppy	Superfluid Behavior of ^{3}He

faculty and publication of research results, a number of which are cited in the references to this chapter. During its first two years of operation, more than 225 visitors came on technical visits to see the facility and discuss technical problems or interaction. In addition, the facility sponsors workshops dealing with microstructures science, engineering, and technology on an annual basis. One of these was the NSF/NRC Airlie Workshop held in November 1978; a separate proceedings of which is available. The Cornell Program on Submicrometer Structures (PROSUS) was established as an industrial affiliates program by the Cornell faculty members associated with the facility and provides a mechanism for close liaison with the industrial community. During the fiscal year 1979–1980, there were eight senior visitors and residents at the facility for a major part of the year. Perhaps the most effective means for discharge of the information-exchange responsibility of the facility's mission will occur as visitors, resident staff, and graduate students interact at the facility and carry new knowledge throughout the country as they take jobs or return to their own institutions.

IV. FUTURE DIRECTIONS FOR NRRFSS

As the facility matures, we expect it to continue to exhibit the broad, interdisciplinary approach to microstructure research that has characterized its initial years. Many of the early results achieved in the facility were outgrowths of programs that had existed previously in the device, materials, and characterization areas. In the future, we expect more impact and results from programs that were initiated after the founding of the facility. These areas include computer research, computer-aided design and circuit layout, work on polymer and inorganic resists, ion and x-ray lithography, and more work in the fabrication of silicon-integrated circuits. A major aspect of the latter work will probably be increased research in dry processing. Many of the programs in fundamental physics are embryonic at this stage and may yield exciting new results as measurements are carried out on structures currently being fabricated.

A major effort is currently underway to serve users outside of Cornell more effectively, and it is expected that this will result in an increase in the usage of the facility by research groups from other institutions. Part of the current work in the facility is aimed at providing sufficient support staff to assist a larger user group and in developing new fabrication tools for microstructures; a prime example of the latter is ion-beam lithography. As these new tools become available, they will open up new research opportunities that can be pursued within the facility by outside users, as well as Cornell faculty. We expect to see prototype machines for beam processing of materials, such as scanned ion-beam systems being constructed within the facility and then made available to users on a broad scale. As this mode of operation increases, we may, therefore, expect to see a flow of new ideas and techniques and instrumentation design for producing and understanding submicron structures from the facility into industry and to other universities.

It is expected that the facility will become more valuable to both university and industrial researchers as it matures. It is well equipped with significant user and core research programs underway and ready to respond further to some of the great opportunities and challenges in this field of microstructures science, engineering, and technology.

REFERENCES

1. G. R. Hanson and B. M. Siegel, H_2 and rare gas field ion source with high angular current, *J. Vac. Sci. Tech.* **16**, 1875–1878 (Nov./Dec. 1979).
2. B. M. Siegel and G. R. Hanson, Gaseous field ion source for submicron fabrication, presented at *Int. Conf. Electron Ion Beam Sci. Technol., 9th, St. Louis, Missouri,* (May 1980).
3. Gary R. Hanson, Metal–Ceramic Bonding Technique, U.S. patent pending.
4. L. Karapiperis and C. A. Lee, 400-Å high aspect-ratio lines produced in polymethyl methacrylate (PMMA) by ion-beam exposure, *Appl. Phys. Lett.* **35**, 395 (1979).
5a. L. Lindhard, M. Scharff, and H. E. Schiott, *Mat-Fys. Medd. Dan. Vidensk. Selsk* **33** (14), (1963).
5b. J. Lindhard, V. Nielsen, M. Scharff, and P. V. Thomsen, *Mat-Fys. Medd. Dan. Vidensk Selsk* **33** (10) (1963).
5c. K. B. Winterborn, P. Sigmund, and K. B. Sanders, *Mat-Fys. Medd. Dan. Vidensk. Selsk.* **37** (14) (1970).

6. D. A. Nelson and A. L. Ruoff, Diamond: An efficient source of soft X-rays for high resolution X-ray lithography, *J. Appl. Phys.* **49**, 5365 (1978).
7. R. A. Buhrman, private communication (1980).
8. D. C. Flanders, X-ray lithography at ~ 100 Å linewidths using X-ray masks fabricated by shadowing techniques, *J. Vac. Sci. Technol.* **16**, 1615 (Nov./Dec. 1979).
9. K. Balasubramanyam, A. L. Ruoff, and K. L. Chopra, Paper Al-7, Meeting of the American Physical Society, New York (March 1980).
10. A. N. Broers, *Appl. Phys. Lett.* **22** 610 (1973).
11. M. S. Isaacson, J. Langmore, and J. Wall, The preparation and observation of biological specimens for the high resolution scanning transmission electron microscope,'' *Scanning Electron Microsc. 1975,* pp. 19–26.
12. M. Isaacson, M. Utlaut, and D. Kopf, Analog computer processing of scanning transmission electron microscope images, *in* ''Topics in Current Physics'' (P. W. Hawkes, ed.), Vol. 13, Computer Processing of Electron Microscope Images, chapter 7. Springer-Verlag, Berlin and New York, 1980.
13. J. Barnard, R.-S. Huang, and Jeffrey Frey, SOS MESFET processing technology for microwave integrated circuits, *IEDM Tech. Digest* p. 281 (1979).
14. J. Barnard, J. Frey, K. F. Lee, and J. F. Gibbons, Micrometer-gate MESFET's on laser-annealed polysilicon, *Electron. Lett.* **16**, 297 (1980).
15. A. Basu and J. M. Ballantyne, Second and higher order waveguide grating filters, II. Experiment, *Appl. Opt.* **18**, 3627 (1979).
16. J. C. Gammel and J. M. Ballantyne, An integrated photoconductive detector and waveguide structure, *Appl. Phys. Lett.* **36**, 149 (1980).
17. J. C. Gammel, G. C. Metze, and J. M. Ballantyne, An epitaxial photoconductive detector for high speed optical detection, *Proc. IEDM* **634** (1979).
18. J. C. Gammel, G. M. Metze, and J. M. Ballantyne, A photoconductive detector for high speed fiber communication, *IEEE Trans. Electron Devices* (in press).
19. C. E. C. Wood, G. Metze, J. Berry, and L. F. Eastman, Complex free carrier profile synthesis by 'Atomic-plane' doping of MBE GaAs, *J. Appl. Phys.* **51**, 383 (1980).
20. B. Chen and C. L. Tang, Nd-glass thin film waveguides: An active medium for Nd thin film laser, *Appl. Phys. Lett.* **28**, 435 (1976).
21. A. L. Ruoff, *Phys. Today* p. 17 (April 1979).
22. A. L. Ruoff and D. A. Nelson, Jr., *Phys. Rev. Lett.* **42**, 383 (1979).
23. C. E. C. Wood, D. DeSimone, and S. Judaprawira, Improved molecular-beam epitaxial GaAs power FET's, *J. Appl. Phys.* **51**, 2074 (1980).
24. C. E. C. Wood, G. Metze, J. Berry, and L. F. Eastman, Complex free-carrier profile synthesis by 'atomic-plane' doping by MBE, *J. Appl. Phys.* **51**, 383 (1980).
25. C. E. C. Wood, L. Rathbun, and H. Ohiwa, On the origin and elimination of macroscopic defects in MBE films, *J. Crystal Growth* **51**, 299–303 (1981).
26. H. Ohno, C. E. C. Wood, L. Rathbun, and L. F. Eastman, GaInAs–GaAlAs structures grown by molecular beam epitaxy (MBE), *J. Appl. Phys.* **52** (1981).
27. S. Tiwari, L. F. Eastman, and L. Rathbun, Physical and material limitations on burn out voltage in GaAs power MESFETs, *IEEE Trans. Electron Devices* **ED-27** (1980).
28. C. L. Huang and T. Van Duzer, Josephson tunneling through locally thinned silicon wafers, *Appl. Phys. Lett.* **25**, 753–756 (1974).
29. D. R. Turner, Electropolishing silicon in hydrofluoric acid solutions, *J. Electrochem. Soc.* **105**, (7), 402–408 (1958).
30. J. P. Langmore, J. Wall, and M. S. Isaacson, The collection of scattered electrons in dark field electron microscopy, *Optik* **38** (4), 335–350 (1973).

31. G. Lehmpfuhl, Convergent beam electron diffraction, *Int. Congr. Electron Microsc.*, *9th*, **3**, 304–315 (1978).
32. P. E. Batson, Damping of bulk plasmons in small aluminum spheres *Solid State Commun.* **34**, 477–480 (1980).
33. G. M. Metze, H. M. Levy, D. W. Woodard, C. E. C. Wood, and L. Eastman, GaAs integrated circuits by selective molecular beam epitaxy, *Conf. Ser. Inst. Phys.* No. 56, Chapter 3.
34. D. L. Hammerstrom and A. Kratzer, Static decentralized load leveling techniques (submitted for publication).

Chapter 5

Limitations of Small Devices
and Large Systems

R. W. KEYES

Thomas J. Watson Research Center
International Business Machines Corporation
Yorktown Heights, New York

185

I. INTRODUCTION

Many chapters in this volume attest to the revolution created by the advent of silicon microelectronics. The impact of semiconductor electronics in so many areas is a direct result of its record of providing ever-increasing information processing power per unit of cost. The steadily decreasing cost per digital operation has been achieved by integration: the fabrication of ever more components—diodes, transistors, capacitors, and resistors—on a single piece, or chip, of silicon. A very primitive approximation asserts that the cost of producing a chip is independent of what devices or circuits are created on it. The number of components that can be fabricated on a single chip has increased from 1 to over 100,000 since 1960 [1]. Information handling elements, including logic, memory, and switching circuits, are based on integrated electronics.

The success of integrated electronics has several origins. Among these, in addition to cost, are low power consumption, high operating speed, and high reliability. The power consumption and high operating speed are rather directly associated with miniaturization, which reduces capacitances as well as the time for electrons to travel through a device. While it may not be apparent that miniaturization is an aid to reliability, such has, indeed, turned out to be the case. The small size of the transistor, and especially of the integrated circuit transistor, has made possible new production and assembly techniques. Integrated electronics places a large number of components on a thin slab of silicon about 5 mm square, a chip, that can be handled as a single unit. Reliability results from the small number of interconnections that must be made by conventional soldering or other bonding techniques, such connections being an important source of failures.

The great reduction in power per device and the greatly increased reliability of solid-state elements combined with lower costs have made possible large increases in the size (as measured by the number of circuits) of electronic systems. The high speed of solid-state circuitry, combined with the increase in size that has become possible, has led to great increases in the power of large electronic computers. It has also led to the development of small and low cost but powerful information processors.

Modern integrated electronics is based on planar technology, which means that devices and circuits are fabricated by operating on a surface of a chip, modifying it chemically by introducing impurities through masks and depositing layers of conducting and insulating substances in selected regions [2]. The central theme of the search for improved digital devices continues to be miniaturization, and the transistor remains the basic

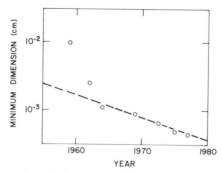

Fig. 1.   Reduction of minimum lithographic dimension with time [1].

device used in electronic switching systems and calculators. A key parameter that measures the status of integrated circuit technology is the minimum lithographically controllable dimension. The historical development of this parameter is shown in Fig. 1, and it can be seen that the minimum dimension has been reduced throughout the past decade by a factor of two about every seven years [1].

An almost equal contribution to the increase in the number of components per chip has come from the increase in the size of the chip. The size of a chip is limited by defects, i.e., the existence of conditions that prevent the fabrication of an operable component in a certain region. A chip must be small enough to make the probability that it contains no defect reasonably large. Improvements in silicon substrates and in processing techniques have steadily reduced the density of defects per unit area and made progress to ever larger chips possible. Chips measuring 0.5×0.5 cm are now common. Both the increase in the chip size and the miniaturization of the components give rise to many difficult and interesting problems.

Thus a rather complex engineering art devoted to the fabrication of microstructures has developed. The art involves the use of photo- and cathodolithography, chemical etches, epitaxial growth, ion implantation, diffusion, oxidation reaction, and many other physical and chemical processes. Examples of microstructures can be found in other chapters.

Two figures will illustrate the progress more quantitatively. Figure 2 shows how the minimum dimension that is used in the commercial fabrication of electronic components has decreased with time and includes a projection of the future of the minimum dimension [1,3,4]. The steady increase of chip size and a projection of its future are also shown. A third factor, called cleverness or ingenuity, has also played a role in the steady increase in levels of integration. A square with edge equal to the minimum

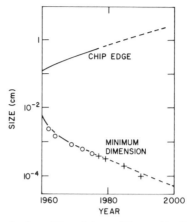

Fig. 2. History and projection of the minimum lithographic dimension and the maximum technologically feasible chip size. The points used to extrapolate minimum dimensions are taken from references [1] (○) and [3] (+). (From Keyes [4].)

lithographic dimension may be regarded as a resolvable element, or *pixel* in the language of display technology. Figure 2 shows that the number of resolvable elements per chip has increased enormously. Cleverness involves making a given component from fewer resolvable elements. As an example, it may be necessary to leave a certain margin of unused area around a component to ensure isolation from other devices. A new isolation technique may reduce this unused area, frequently, however, at the expense of an additional processing step.

Miniaturization also plays a large role in the steady decrease of the energy utilized per switching operation since the introduction of transistors into computers. The product of circuit power and delay time tends to be more characteristic of technology at any given time than either quantity separately. The energy is determined by the voltages and the capacitances and can be supplied slowly or rapidly. The trend of this parameter, the product of power per circuit and logic delay (*pt*), is shown in Fig. 3 [4,5]. The energy dissipated has decreased by about four orders of magnitude since 1957. The sources of this decrease may be very roughly assigned as follows: one order of magnitude to decrease in voltage (we have suggested that the voltage is approaching a lower limit and cannot be decreased much further [6]); two orders of magnitude to miniaturization; one order of magnitude to improved designs, such as replacing a $p-n$ junction area with a semiconductor–insulator interface. Historically, the power–delay product of integrated logic chips has decreased nearly in proportion to the area per circuit. Thus, it can be expected that circuits will be operated

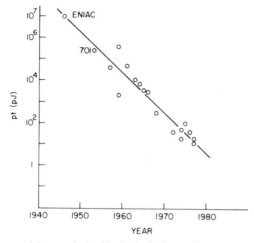

Fig. 3. The power–delay product of logic, including early vacuum tube as well as transistorized computers [5]. (From Keyes [4].)

with less power. Nevertheless, because of the projected rapid increase in number of circuits per chip, it is desirable to increase the density of power dissipation. Difficulties in removing heat directly degrade circuit speed because of the power–speed trade-off.

II. PHYSICAL NATURE OF INFORMATION PROCESSING

Electronic information processing is based on the representation of information in physical form. Memories, communication links, and logical processors deal with pulses of current and charge on capacitors and other entities that are describable by the concepts of physics. Information is contained in the physical state of a structure. Although the physical state of a structure or body of matter or the fields in a region of space may be completely well defined in principle, there is uncertainty associated with the determination of the state or of the properties of the structure or material involved and, therefore, an uncertainty in the information that is represented.

Some uncertainty derives from the basic physics of the phenomena used to represent information, such as the quantization of electromagnetic waves and the thermal agitation present in all material systems. The course of events in physical systems can only be described probabilistically. Quantities such as the time when an excited electronic system will decay and emit a photon or the number of electrons in a material body whose energy exceeds a specified value are not known exactly but their probability distribution is known. These kinds of uncertainties make precise prediction of the future behavior of a system impossible. The uncertainty associated with lack of perfect microscopic knowledge of a system is encompassed in the concept of noise and is familiar in electrical devices as Johnson noise, generation–recombination noise, and shot noise, which can be described by quantitative theory.

A second kind of uncertainty is caused by fluctuations whose source is in the environment, such as the induction of unwanted signals on conductors by electromagnetic fields that are changing in ways that have nothing to do with the information being transmitted through the conductors. Such uncertainties also can be described as noise.

Still another kind of uncertainty derives from the lack of complete and perfect knowledge of the physical characteristics of any manufactured device. Quantities such as the dimensions of a structure and the exact composition of a material body and its dependence on position cannot be controlled or determined with unlimited precision. This lack of knowledge of the fine details of the system being measured introduces an additional uncertainty into the information content of a device or of the measurement of a signal.

The effects of uncertainty in information handling can be overcome by the use of large signals. This is well known in communication, and the effects of signal power on communication capacity in the presence of noise were given quantitative expression in a historic work by Shannon [7]. Although the considerations involved in digital devices are somewhat different in detail and are more likely to involve uncertainties in component values, the result that large signals are needed still applies. For example, if the threshold voltage of a field-effect transistor is uncertain by an amount ΔV, then a large signal voltage compared to ΔV is required to ensure the changing of the state of the transistor. The signal power is eventually dissipated and turned into heat, which must then be removed from the system. The removal of heat from systems is, however, increasingly difficult as the systems become miniaturized. The limits on the performance of a large compact information device that are set by the problem of removing the heat produced are directly connected to the absence of perfect knowledge of the properties of the device.

Computer operations have great depth in the sense that information may be subjected to a very large number of operations during a computation [8]. Even a small error in each step can cause complete loss of information after a large number of steps. Digital representation is used to avoid this loss of information. The degradation of information is pre-

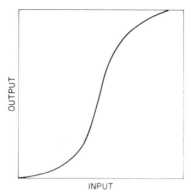

Fig. 4. A nonlinear response that restores binary signals toward their values at the input [9].

vented by restandardizing each digit at each step. Thus, if information is represented by a series of ones and zeros, then a zero is restored to zero signal level while it is still recognizable as a zero, and similarly for a one. An error in signal level caused by variation in a parameter that characterizes some component of 10 or 20% or more may be accommodated. Digital representation makes arbitrary accuracy possible simply by extension of the number of bits used in the representation. The role of standardization at each step is crucial, and the response that a physical system must have to standardize signals is shown in Fig. 4 [9].

Thus the electrical devices of a modern computer perform elementary logical operations on binary integers. A circuit receives inputs representing digital information from other circuits, constructs a standardized electrical representation of the result of some logical operation on the inputs, and transmits the standardized output signal to one or more other circuits.

It is apparent from Fig. 4 that a circuit that provides a standardized output must have a nonlinear response. The degree of nonlinearity needed to achieve a useful response of the form of Fig. 4 is increased by the variability of components. The physical basis of electrical nonlinearity can be seen as follows: The application of an electric potential to an electronic device changes the energy of the electrons on, say, the n-side of a $p–n$ junction with respect to the energy of other electrons (those on the p-side of the junction). However, electrons are dispersed in energy by an amount of approximately kT by thermal agitation, where k is the Boltzmann constant and T absolute temperature. Electrical voltages that are small relative to kT/q are a small perturbation of the electrons and produce only linear effects (q is the charge of an electron; $kT/q = 0.025$ V at $T = 300$ K). Nonlinear effects can be produced by voltages that are large relative to kT/q.

This scale of nonlinearity is most perfectly exemplified by the ideal $p-n$ junction in which the current depends on the voltage as

$$i \propto \exp(qV/kT) - 1. \tag{1}$$

The junction characteristic as expressed by Eq. (1) is a practical limit to electrical nonlinearity. A similar scale even seems to be applicable in biology; i.e., neuron voltages are a few times kT/q. Various lines of thought agree in establishing kT/q as a practical minimum voltage scale for the production of nonlinear electrical effects. The word *scale* implies that the attainment of the very large nonlinearities needed for reliable logical operation in the presence of such disturbing influences as cross talk, environmental fluctuations, and component variability will require that voltages a great many times kT/q be used in real circuits.

A specific example of the relation between signal voltage and thermal voltage was obtained in a study of the threshold characteristics of a field-effect transistor as a function of temperature [10]. Examples are shown in Fig. 5, and it is seen that the voltage necessary to convert the transistor from its off to its on state increases in proportion to the temperature.

The voltage requirements of digital electronics have been stressed, since energy is dissipated in the charging and discharging of capacitors and in resistors, and the dissipation will increase with increasing voltage.

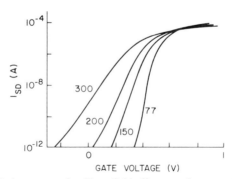

Fig. 5. Source–drain current of a silicon field-effect transistor as a function of gate voltage and temperature, illustrating the sharpening of characteristics as the temperature is lowered [10]. The curves are labeled with the temperature in kelvins. (From R. W. Keyes, *Science* **195**, 1230–1235 (1977). Copyright 1977 by the American Association for the Advancement of Science.)

III. DEVICES FOR LSI

A good deal of attention has been focused on devices for large-scale, high-speed integrated circuits. Until now silicon bipolar transistors have dominated high-speed, solid-state electronics. Recently, however, increasing attention has been attracted to gallium arsenide and silicon MESFETs (Metal–Semiconductor Field-Effect Transistors) for high-speed circuits [11,12]. Even insulated gate FETs on silicon, which have been the mainstay of low-cost semiconductor memory and have made the highest levels of integration possible, cannot be excluded as a contender for high-speed logic [13].

One reason for the increasing interest in new high-speed technologies is that bipolar technology becomes increasingly difficult to practice as miniaturization progresses and dimensions decrease. Typical forms of the semiconductor devices in question are illustrated in Fig. 6. It can be seen that the fabrication of bipolar transistors requires the production of complex patterns of dopants in the semiconductor. As the MESFET is probably the least familiar device, the Appendix is devoted to a short description of some of its characteristics.

The relative advantage of various transistors can be seen by considering

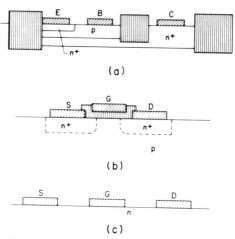

(a)

(b)

(c)

Fig. 6. The physical forms of various types of transistors that must be considered for high-speed logic: (a) bipolar; (b) IGFET; (c) MESFET. The electrode symbols mean: E, emitter; B, base; C, collector; S, source; D, drain; G, gate. The vertical shading denotes insulators; the diagonal shading denotes contacting metals. (Courtesy of Electronic Conventions, Inc.)

that their basic function is to control current to charge and discharge capacitances. In integrated logic much of this capacitance is in the wires that interconnect devices on a chip. Thus we may write the delay time for a switching operation as

$$t_d = \tfrac{1}{2}(fC_d + C_w)\,\Delta V/i, \tag{2}$$

where C_d is the capacitance of a device, f the fan out or number of other devices that must be driven, C_w the wire capacitance, ΔV the voltage change, and i the current supplied by the driving transistor. The factor $\tfrac{1}{2}$ occurs because the switching of the next stage begins after half of the voltage transition has taken place.

Thus t_d can be decreased by increasing i. The current in a bipolar transistor can be increased by adjusting resistances and bias voltages in the circuit; large currents can be carried by a small transistor. The current in the various field-effect transistors is more limited. For example, in a MESFET, the current cannot exceed that which can be carried by the electrons in the conductive layer moving at their maximum or limiting velocity. The design current of an FET can only be increased by increasing its width—within the constraints of a given processing technology, e.g., doping level, oxide thickness, and lithographic capability. Increasing the width of an FET, however, increases the size of the cell that is occupied by a logic gate.

Figure 7 shows how the area of a cell containing one logic gate is used for devices and wiring. Obviously, the total amount of wire needed de-

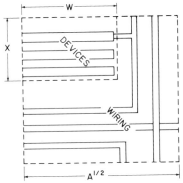

Fig. 7. A cell containing a logic gate with an area utilized by devices, proportional to transistor width, and an area used for wiring, proportional to total wire length. The area of the cell is A.

pends on the size of the cell; the larger the cell size, the longer the wires that interconnect them must be. On the other hand, the longer the wires, the greater the area required for wiring. There is a type of positive feedback: Since increasing the size of the devices also expands the cell and increases the total wire length, a point of diminishing returns from increases in device size is reached.

This situation can be modeled quantitatively. The area of the part of the cell that contains devices is proportional to the width of the devices and the area of the wiring is proportional to the length of the wires. The average length is proportional to the cell edge, the square root of the cell area. Thus, it is possible to write for the cell area A and the average wire length L,

$$A = XW + fYL, \tag{3}$$

$$L = mA^{1/2}, \tag{4}$$

where W is the width of an FET, X and Y are constants with the dimensions of length, m is the average wire length measured in circuit pitches, and f is another dimensionless constant that describes the fan out, or the number of wires generated by each circuit. Equations (3) and (4) represent a quadratic equation that can be solved for A and L.

Figure 8 shows an example of the dependence of delay time on power and area as calculated from Eqs. (2)–(4). Substantial gains in speed can

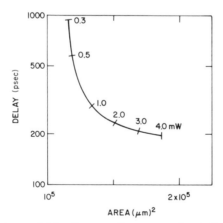

Fig. 8. The dependence of delay time on cell area according to the model of Fig. 7 for a particular choice of parameters under 1-μm technology.

be attained by increasing the area per gate and the power, which is proportional to the current, when the MESFETs are small, but beyond a certain point the accompanying increase in wiring length largely counter-
balances the advantage gained by the increased transistor width. It is not
surprising that FETs have found their principal application in low-
current, low-power circuits, while bipolar transistors have dominated applications in which high power is used to attain high speed.

IV. LIMITS TO MINIATURIZATION

A. Lithography

Continued miniaturization depends on improvements in lithography,
the art of producing patterns in films that can be used as masks in processing, for example, to define patterns on a substrate where material may
be altered by etching or by introduction of impurities. Lithography has received a substantial amount of attention as a limit to miniaturization
[14,15]. The most common form of lithography is photolithography in
which exposure to light changes the properties of a film of a photosensitive compound, allowing it to be selectively removed for purposes of
masking. Exposure to light cannot, in practice, produce patterns with
minimum dimensions much less than the wavelength of the exposing light.
Thus, not too many years ago, dimensions of 1 or 0.5 μm were viewed as
the limiting least count of microstructures.

The exposure of resist with electron beams has, however, emerged as a
method that can produce much finer structures [16]. In fact, there appears
to be no significant fundamental limit to the resolution that can be
achieved with electron beams; single atoms can be seen with the electron
microscope [17]. There are, however, practical limits. A very large
number of resolution elements must be exposed to form a microstructure.
For example, a chip 3 mm square is not large in light of modern technology, but to expose it with 1-μm resolution implies exposure of 10^7 elements. A feasible production process must expose each of the 10^7 elements in a very short time. The highest possible beam current is desirable.

Thus a presently accepted view of the limits to electron-beam
lithography runs as follows [18]. Spherical aberration in electron lenses
spreads the electron beam by an amount proportional to the cube of the
incidence angle α accepted at the target (see Fig. 9). The diameter of the
exposed spot is at least

$$D = C\alpha^3/2, \qquad (5)$$

where C is a constant that characterizes the spherical aberration of the electron lens. The current density in the beam is limited by the brightness B of the source (in amperes per unit area per steradian). The current in the beam is proportional to the solid angle accepted or to α^2, and thus large currents and small dimensions conflict according to Eq. (5). The electrons in the beam arrive randomly in time and are distributed in any specified time t according to Poisson's law. To ensure that statistical fluctuations do not leave any resolution element underexposed, it is necessary that the average exposure of an element exceed some minimum number of electrons N. For example, if $N = 20$, then there is probability of 0.006 that an exposure element receives only 10 electrons. Asking that the probability of an exposure less than half the average be smaller than 10^{-14} requires that N be something like 200. These limits on the beam current and on N imply that there is a minimum exposure time per spot, which depends on the spot diameter. The relation between the exposure time and D is shown in Fig. 9. As the cost of an electron-beam exposure tool must be spread over a large number of components, Fig. 9 implies an economic limit on

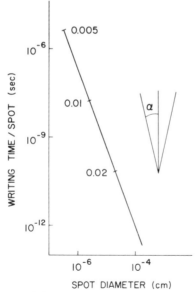

Fig. 9. Minimum time needed to expose a picture element of an electron resist with a focused electron beam as a function of element diameter [18]. The limit is calculated with the following parameters, which are typical of modern electron-beam systems: spherical aberration $C = 5$ cm; source brightness $B = 10^6$ A/cm² sr; minimum number of electrons per spot $N = 200$. The values of α, the beam angle accepted in Eq. (5), are indicated on the limiting line. (From R. W. Keyes, *Science* **195**, 1230–1235 (1977). Copyright 1977 by the American Association for the Advancement of Science.)

the applicability of electron lithography. Only very rough values of the parameters involved are needed to quantify this limit because of the very strong dependence on D. For example, taking the cost of the exposure system as $100/hr and the value of the exposed silicon as $10/cm^2, an exposure rate $R = 3 \times 10^{-3}$ cm^2/sec is required, and it is found that $D > 5 \times 10^{-6}$ cm.

It would be a mistake to regard such a limit on electron-beam exposure technology as fundamental, however. One possible escape might be found in the suggestion that it is possible to eliminate completely the effects of spherical aberration [19]. Clearly, the invention of brighter electron sources could directly affect the exposure time. The potential of electron-beam lithography is far from exhausted.

There is, though, another very important limit to the resolution of electron-beam lithography as practiced today. Electrons pass through the resist, are scattered in the underlying silicon, and may emerge some distance away from the incident beam to expose the resist there [16]. The range of a typical 25-kV electron in silicon is about 3 μm, so there is a nonnegligible degradation of resolution by this backscattering effect. The effect can be diminished by reduction of the energy of the exposing electrons. Decreasing the electron energy, however, increases the chromatic aberration of the electron lenses, introducing another source of spreading of the spot.

The large-angle scattering that causes electrons to be returned from the substrate to the photoresist could be avoided by exposing with more massive particles, i.e., protons or other ions. The low brightness of ion sources has prevented the useful exposure of resists with ions to date. However, the development of more intense ion sources may be another path to pushing back the present limits on lithography [20,21].

In x-ray lithography an attempt is made to take advantage of the very high resolution that is possible in principle in writing with electron beams without being restrained by the long times that are needed to expose a chip by sequential scanning [14,15,22]. A high-resolution mask made by writing with an electron beam can be used to expose photoresist on a silicon substrate with x rays; the short wavelength of the x radiation, 1 to 100 Å, preserves the high resolution of the mask. Since a mask can be used to expose a large number of substrates, a long time and low current can be used to achieve high resolution in the preparation of the mask. The resolution attainable with x rays is limited by a different effect: a secondary electron is emitted when an x-ray photon is absorbed, and the secondary electron has a range of one to several hundred angstroms and exposes the resist in an area with this radius.

Thus, it must be reemphasized that the limits of particle-beam lithography just described are in no way fundamental; they involve practical and economic considerations. Indeed, fabrication of metal conductors 80 Å wide has been reported [23]. One can be confident that the march of lithography to smaller dimensions will continue. Physical limits to miniaturization must be sought elsewhere.

B. Hot Electrons and Breakdown

It was argued in Section II that voltages cannot be reduced indefinitely but must be large in comparison with the thermal voltage and voltage "noise" due to process variability. Nevertheless, all dimensions of a structure, including such internal device dimensions as base widths in bipolar transistors and the thicknesses of the depletion layers between p- and n-type regions of silicon, are decreased as miniaturization advances. Electric fields and high-field phenomena such as hot electrons and dielectric breakdown grow in importance with decreasing dimensions and form a limit to miniaturization.

A quantitative version of one such limit has been formulated [24]. Consider the field-effect transistor shown in Fig. 10. The depletion regions associated with the p–n junctions of the source and drain electrodes are shown. The source–drain distance must be greater than the sum of the widths of the depletion layers in order for the gate to exercise control of the conductance along the surface. The width of the depletion layers can be reduced by increasing the doping level of the substrate silicon. The source and drain can then be placed closer together and the transistor made smaller. However, the electric fields in the depletion layers will be increased at any given applied voltage, and the breakdown voltage of the transistor will be decreased. Further, there is an electric field in the oxide insulator when a conductive surface is induced by a voltage applied to the gate. The increase in doping leads to an increase in electric field in

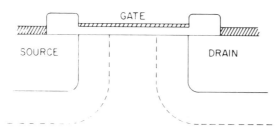

Fig. 10. Structure of an insulated gate field-effect transistor showing the extent of the depletion regions associated with the source and drain junctions.

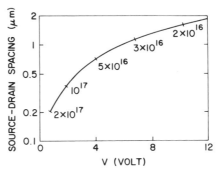

Fig. 11. An example of the limits imposed on a metal oxide semiconductor field-effect transistor as shown in Fig. 10 by breakdown in the SiO_2 insulator [24]. Heavy doping reduces the widths of the depletion layers illustrated in Fig. 10; the impurity concentrations needed to achieve particular source–drain spacings are indicated. Breakdown in the oxide then limits the supply voltage of a simple inverter to the value shown. (From R. W. Keyes, *Science*, **195**, 1230–1235 (1977). Copyright 1977 by the American Association for the Advancement of Science.)

the oxide. Thus both junction breakdown and oxide breakdown limit the miniaturization of FETs. Since breakdown of silicon and of SiO_2 are well-studied subjects, it is possible to calculate for each voltage the maximum doping level that can be used and the smallest permissible source–drain separation. A calculation of this type is shown in Fig. 11 [24]. The results presented show that breakdown in the oxide insulator is the limiting factor in the miniaturization of FETs. However, curvature of a junction and nonuniform doping profiles affect the breakdown voltage but are difficult to take into account quantitatively. It has also been suggested that breakdown at the drain–substrate junction is the more serious limitation.

Related models can be developed for bipolar transistors. The basic requirement is that the base shall not be completely depleted or suffer punchthrough and that the junctions shall not break down. The punchthrough is controlled by heavy doping, which, however, decreases the breakdown voltage of the junctions and the voltage at which the transistor can operate.

The electron temperature, a rough measure of the average electron energy, rises above the lattice or thermal equilibrium temperature at electric fields much smaller than those needed to cause avalanche breakdown. The electrons become hot, and some of the hot electrons have enough energy to pass from the silicon into the SiO_2 insulator on the surface and become trapped there, simulating a potential applied to the gate and changing the properties of the surface [25]. Such effects are noticeable in both field-effect and bipolar transistors. In FETs hot electrons produced by electric fields in the channel at the surface change the threshold voltage when they escape into the insulating SiO_2. Hot electrons that are produced in the region close to the intersection of a $p-n$ junction with the silicon surface increase the leakage current across the junction along the surface when they escape into the SiO_2 and degrade the performance of bipolar transistors. Quantitative studies of such effects in FETs are available and show that undesirable changes in characteristics occur rapidly if the fields in the channel are only slightly greater than 10^4 V/cm [25a]. Fields in the depletion region under the channel and at the drain junction also produce hot electrons. Although more experimental information is needed, that which is available suggests that these hot electron effects will limit reduction of the FET length—the source–drain separation—to something like $\frac{1}{4}\,\mu$m. Information concerning the hot electron degradation of $p-n$ junctions is even more scanty.

C. Small Number Fluctuations

The number of dopant atoms contained in a device can become small if the device dimensions are small. Since the dopant atoms are randomly placed in the semiconductor, fluctuations of the number of dopant atoms in a region can become significant in miniaturized devices. This effect is important in depletion regions of semiconductor devices, regions from which the mobile charges have been depleted. The variation of electric potential across such regions is controlled by Poisson's equation, and a fluctuation in the charge implies a fluctuation in the thickness of the depleted region or in the voltage drop across it.

Shockley introduced a cube approximation to treat this problem: the depleted region is visualized as composed of cubes whose edges are equal to the thickness of the region [26]. The variation of potential is determined by Poisson's equation separately for each cube, depending on the average charge in the cube. If the thickness of the depleted region is D and the average concentration of dopant atoms is N, then the average number of dopant atoms in a cube is

$$M = ND^3, \tag{6}$$

where N and D are also connected by the integration of Poisson's equation

$$NeD^2 = 2KV. \tag{7}$$

Here V is the potential drop across the depleted layer, K the permittivity of the semiconductor, and e the electronic charge. Because of Eq. (7) the doping level increases as miniaturization progresses since D must be decreased as dimensions decrease. By eliminating N from Eqs. (6) and (7), it is found that

$$M = 2KVD/e. \qquad (8)$$

Thus M decreases and the importance of fluctuations increases as dimensions are reduced.

The standard deviation of the number of atoms in a cube is $M^{1/2}$. Since there are many cubes in a single device (D is usually less than the transverse dimensions of a device) and many more on a chip containing hundreds or thousands of devices, fluctuations of several standard deviations are likely to be encountered.

Fluctuations in the relative magnitude of the device parameters, the voltage, or layer thickness will be of the order of $M^{-1/2}$. As examples of the magnitudes involved, if $V = 2$ V and $D = 1$ μm, then $M = 2500$ and $M^{-1/2} = 0.02$. If $D = 0.1$ μm, then $M^{-1/2} = 0.06$.

In the case treated by Shockley, i.e., the breakdown of a $p-n$ junction, breakdown is determined by the narrowest part of the depleted region, the smallest cube. It was found that fluctuations decrease the breakdown voltage of a silicon $p-n$ junction by about 0.7 V. In another application of these ideas, the threshold voltage of field-effect transistors is determined by the charge in a depleted layer [27]. The threshold voltage is a property determined by an average over many cubes, and the fluctuations are reduced by this further averaging. Depending on the parameters of the FET, one may expect to find a variability of thresholds ranging from 0.01 to almost 0.1 V on an integrated circuit chip [28].

V. LARGE SYSTEMS

Economically, the most important impact of LSI devices will almost certainly be as individual chips performing complex functions in the machinery of our civilization, in such things as automobiles, typewriters, washing machines, telephones, pinball games, and cash registers. However, LSI is also invading the technology of large computers, which are needed to handle problems of great scope, such as calculational chemistry, numerical weather forecasting, engineering simulation, and the management of very large data sets and switching systems. These large systems have the following characteristics that are not relevant to the use of LSI devices as isolated functional units:

(1) They are assembled from many chips.
(2) A great many interconnections among the chips must be provided.
(3) There are many different kinds of chips or part numbers.

These characteristics add up to great complexity. The design of a single LSI chip is difficult. In large systems the chip designs must interact and influence one another. A certain synchronism of the operation of the different constituents of the system must be ensured. Packages, substrates on which the chips are mounted, must contain a large assemblage of interconnections.

A. Chips

General principles that penetrate the complexity of large systems are difficult to find. There are, however, statistical features that seem to characterize many such systems and their components. Although it will be assumed that the statistical averages describe various properties of the large system, it must be understood that large systems are designed for differing purposes and some may not conform to the averages. The statistical features used here are drawn from the properties of large, general-purpose computers.

The most familiar statistical property of large systems is the relation between the number of logical circuits or gates that are contained in a subsystem, such as a chip or some kind of multichip carrier, and the number of connections that must be made to the subsystem [29]. The relation has become known as Rent's rule and has the form

$$n_c = BN^j, \tag{9}$$

where N is the number of circuits on a unit and n_c the number of signal connections that must be provided; B and j are constants whose values differ among authors but here $B = 2$ and $j = \frac{2}{3}$ will be used. Available data do not extend to entities requiring more than 10^3 connections, however, and the view that Eq. (9) does not apply to functional units with $N > 1.1 \times 10^4$ gates and that 10^3 is the maximum number of connections needed will be adopted. The resulting form of the n_c versus N curve is shown in Fig. 12. Equation (9) is empirically based and represents what system designers have found necessary to ensure good circuit utilization.

It is by no means obvious, however, that 10^3 is a maximum value for the number of interconnections required by any functional subsystem, and in-

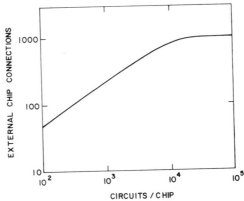

Fig. 12. The number of signal-carrying connections that must be made to a logic chip as a function of the level of integration [29]. The number of connections is proportional to the $\frac{2}{3}$ power of the number of circuits at low levels of integration but is not extrapolated beyond 1000 connections. (From Keyes [4].)

troducing such an assumption is a response to the scantiness of available statistical information on electronic systems. Indeed, one may find examples of vastly larger numbers of interconnections between subsystems in the human nervous system. The optic nerves, which connect an eye to the brain, each contain about 10^6 fibers, and the corpus callosum, which connects the two halves of the brain, contains 2×10^8 fibers. These numbers lend a certain credence to Rent's rule as a law of nature. If one, admittedly rather speculatively, identifies N with 10^8, the number of optical receptors in the eye, and as somewhere between 10^{10} neurons and 10^{14} synapses in half of the brain, it turns out that Eq. (9) is applicable to these aspects of the human nervous system, as shown in Fig. 13. Equation (9) should also be contrasted with semiconductor memory chips, where the number of connections grows only logarithmically with number of bits.

Another statistical property is the average line length on a chip. This quantity is conveniently measured in dimensionless circuit pitch. The circuit pitch is defined as the distance between adjacent circuits if they were arranged in a square array on the chip. In other words, the pitch on a chip of area A containing N circuits is $(A/N)^{1/2}$. The line length as measured in pitches is found to increase with N, as shown in Fig. 14, where the line suggests that the dimensionless on-chip connection length increases approximately as the $\frac{1}{3}$ power of N.

One way of looking at the result shown in Fig. 14 and giving it some plausibility is as follows. A certain flexibility of choice is needed to inter-

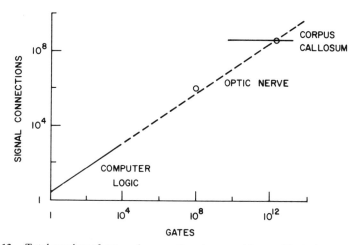

Fig. 13. Total number of external connections to assemblages of logical processing elements for computer logic and for elements of the human nervous system. The number of logic gates that should be ascribed to the eye and to the brain can only be estimated very roughly. (Courtesy of Electronic Conventions, Inc.)

connect individual logic gates to form a computer, which flexibility might be measured by the number of possible interconnection patterns that satisfy the constraints imposed by the physical nature of the system. One such constraint is wire length, which must be limited on a chip because a chip of finite size can only accommodate a fixed total wire length and because wire length must be traded against manufacturing yield. A limited number of external connections, by limiting the number of interconnections that can cross a chip boundary, is another physical constraint. It is seen from Eq. (9) that the number of off-chip connections per circuit decreases as the level of integration N increases, so that the number of interconnection patterns that can be provided on the chip carriers decreases

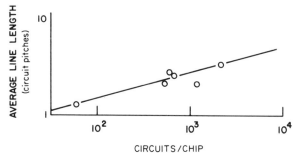

Fig. 14. Dependence of average on-chip interconnection length, measured in circuit-to-circuit distance, on level of integration [29]. (Courtesy of Electronic Conventions, Inc.)

(for a fixed number of logic gates). More possibilities must be made available on the chip to provide the same total number of patterns from which the logic implementation may be chosen. A longer average wire length on the chip increases the number of possible on-chip patterns.

Another problem, created by the inaccessibility of internal chip connections to external contact, is that of testing. The ratio of the number of signal points available for probing, i.e., the off-chip connections, to the number of logic gates decreases with increasing N; less information per circuit is available for diagnosis and test. This feature must be taken into account in integrated circuit design and special provision to ensure testability included. Engineering changes, i.e., modifications of chip design made after the chip has left the design stage, are also affected. They cannot be made by simply moving a wire or two but require the fabrication of new wiring masks for the lithographic process.

The parameters illustrated by Figs. 12 and 14, for which average values are available, are important to the performance, or speed, of a system. The long wires on a chip represent capacitive loads that slow the operation of circuits. The off-chip connections are burdensome in several ways. They are generally long enough that they must be thought of as transmission lines. The low impedance of transmission lines on a package means that large currents must be used to produce the required signal voltages on them; more power is needed to drive them than to operate on-chip circuits. Even though off-chip drivers may be only a small fraction of the total circuit count of a chip, they can account for up to half of the power consumed.

In addition, switching one of these off-chip drivers means that the total current supplied to a chip is changed. If a chip contains a large number of off-chip connections, there is a certain probability that many of the off-chip circuits will be switched simultaneously, producing a large change in the current flowing to the chip in a small time interval. The rapid changes of current in the inductance of the power supply cause disturbances of the supply voltage that can affect the operation of the chip.

The long delays that signals sent from chip to chip encounter also slow the system. The importance of these delays depends on the relative frequency with which the destination of a signal is found on and off of the chip; the more logic gates on a chip, the more often the output of a logic gate will be sent to another gate on the same chip. A chip containing only one circuit necessarily transmits each result to another chip. In fact, one expects that the frequency with which signals are sent off of a chip must bear some relation to the ratio of output connections to circuits on the chip, which, as is seen from Eq. (9), is proportional to $N^{-1/3}$.

Questions must be raised as to what extent the relations shown in Figs.

12 and 14 are fundamental features of high-speed computers and to what extent they represent requirements that might be overcome by increased ingenuity? It is clear that the limitations represented can be relaxed by sacrificing performance. For example, fewer off-chip connections can be used by time multiplexing signals passing through a connection. Some circuits would be needed for the multiplexing function, decreasing the logical processing power of the chip, and signals would be delayed by the need to wait for an opportunity to be transmitted through a connection. It is likely that the number of off-chip connections used could be reduced to an unknown extent by increased investment in design. The general principles that would permit a quantitative or qualitative grasp of trade-offs of this type are lacking.

When a large system is constructed from integrated circuit chips, many functionally different chips are used. A chip type is known as a part number, and the management of large numbers of part numbers is a problem for system designers and manufactureres [30]. The problem did not exist at the chip level in the pre-LSI era, when systems were formed from separate transistors. Apparently it will be mitigated by the use of very high levels of integration, when a system might be contained on a very few chips. The number of separate part numbers seems to peak at levels of integration of 10^2 to 10^3 circuits/chip, where many hundreds of different chips may be used in a large computer [30].

One approach to the provision of many part numbers is through a masterslice [31]. The final processing step that makes one chip distinct from another consists of fabrication of a layer of wiring by lithography. The function of the chip is then controlled by the interconnections provided by

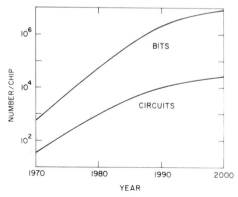

Fig. 15.   An extrapolation of the levels of integration of main memory and logic chips, tempering the capabilities presented in Fig. 2 with the difficulties discussed in Section V. (From Keyes [4].)

this final wiring layer. Since no additional chip space is needed, circuit density and operating speed tend to be preserved.

The collection of problems just recited will hinder the application of LSI to chips designed for large systems to levels below those that will be otherwise technically feasible and that will no doubt be encountered in other applications, such as isolated microprocessors. It is difficult to quantify the various difficulties listed, so that our estimate of the future of LSI in large systems, shown in Fig. 15, contains a large ingredient of guesswork.

B. Packaging

Chips are mounted on larger carriers, frequently made of ceramics, when they are assembled into a larger system. The chip carriers and the substrates, or boards, on which they are mounted provide for mechanical and electrical attachment of the chip and contain the wiring matrix that interconnects the chips and the power distribution conductors [32].

The art of mounting the chips on a substrate in such a way that they can be interconnected and cooled is called *packaging*. Packaging problems comprise one of the most severe limitations on the performance of modern, high-speed, integrated logic. The many functions that must be taken into account in the design of a package are the following:

(1) provide mechanical support and attachment;
(2) provide electrical connection to chip;
(3) transform chip contact dimensions to mechanically pluggable dimensions;
(4) contain wiring matrix for chip interconnections;
(5) contain power distribution network;
(6) receive heat from chip and deliver it to a fluid; and
(7) protect the semiconductor from the environment.

Simple mechanics may limit the closeness with which chips may be placed on a large substrate through requirements such as making the chip carrier conveniently replaceable for fault correction or design changes. Off-chip connections are large as the miniaturization of wire bonds and solder joints has not kept pace with the miniaturization of semiconductor components. Equation (9) can be combined with the package area values derived from a thermal model in Section VI.A to construct Fig. 16, which gives a measure of the miniaturization that will be required of interconnections to maintain present rates of progress in the future. The problem of alloting physical space to interconnections exists on the package as

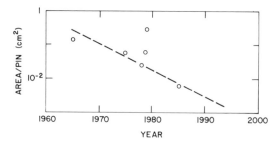

Fig. 16. Trends in the area required per chip–multichip carrier connection.

well as on the chip. The space on the package must also allow for packaging interconnections and mechanical assembly.

Heat dissipation can also limit the density of chips. It is desirable to place chips as closely together as possible to minimize the chip-to-chip transit times in large systems. Since each chip dissipates power to heat, the power density becomes very high and the provision of cooling means limits the density at which chips can be assembled on a package. The speed of on-chip circuits can be increased by supplying them with more power, but since this affects the cooling requirements imposed on the package, one must be traded against the other.

Advances in cooling technology seem imminent, however, and Fig. 17

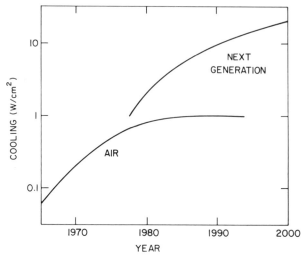

Fig. 17. A projection of cooling capabilities for logic chip packages. Forced air cooling is probably limited to about 1 W/cm². New cooling technologies, in which heat is transferred to a liquid without the intermediary of air, are emerging [33,34], and it has been assumed that they will extend cooling capability to 20 W/cm². (From Keyes [4].)

shows a history and forecast of cooling of large computers. Circuits are conventionally cooled by transfer of heat from a semiconductor or semi-conductor module to air. However, air cooling is limited to a power density of about 1 W/cm², as shown in Fig. 17. Emerging technologies, in which heat is transferred directly to a liquid without the intermediary of air [33,34], break through the 1-W/cm² limit and are probably limited to something like the maximum rate of heat transfer to boiling freon, about 20 W/cm², as also illustrated in Fig. 17.

VI. PERFORMANCE

A. Modeling System Performance

This section is devoted to combining the projections of basic technology (Figs. 2, 17, and 18) and the guess at levels of integration (Fig. 15) with the statistical features discussed in the preceding section to predict performance of large computers. (Figure 18 is a projection of the power–delay product of logic, an extension of Fig. 3 into the future.) A difficulty must be mentioned here: the difference between average values

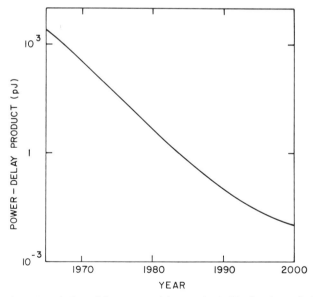

Fig. 18. An extrapolation of the power–delay product of logic, shown in Fig. 3, to the year 2000. The extrapolation is drawn to approach a limit of 0.01 pJ. (From Keyes [4].)

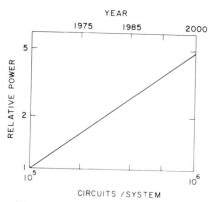

Fig. 19. A projection of the number of circuits per computing system and the effect of the increase in circuits on the relative computational power of the system. (From Keyes [4].)

and worst cases. It may be hoped that the long delays, for example, chip-to-chip delays, will add with shorter ones in such a way that an average value will determine system performance. However, it may turn out that many long delays will combine with one another in a few cases, and a computer cycle will have to be long enough to allow such a worst case series of operations to be completed. This eventually can be contained to a certain extent by identifying extra long paths and making physical or logical design changes to eliminate them, but one cannot be sure how closely average values will be attained. In the absence of better information, this slanting of times toward higher values than the average will be taken into account only by using values on the pessimistic side of the range for parameters whose values is somewhat uncertain.

It is anticipated that system learning will occur and that it will be possible to utilize ever-increasing numbers of circuits in a single system to speed the instruction stream. The relative power of a system is assumed to grow slightly slower than proportionately to the system circuit count, as shown in Fig. 19, and the circuit count is allowed to grow from 2×10^5 in 1975 to 10^6 in the year 2000. The meaning of this in the model of performance is that the time per instruction is assumed to decrease from 60 to 20 packaged logic delays during the period.

The implications of this system assumption for the chip connection problem are shown in Fig. 20. The number of logic chips decreases because of the rapid increase in level of integration. The number of chip connections also decreases somewhat. The prevention of a substantial growth in total number of chip connections is regarded as an objective of design in these projections and implies that a much lower level of integration of the logic chips would be unacceptable.

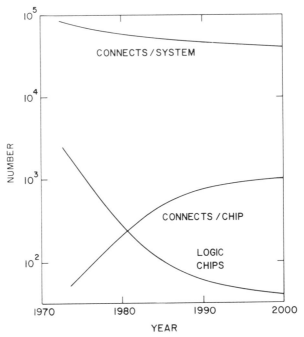

Fig. 20. The provision of chip-to-package connections is regarded as a costly element of computer system manufacture. Thus, one of the objectives of the projection of the integration level of logic chips shown in Fig. 15 was to decrease the total number of off-chip connections used in light of the growth of system size shown in Fig. 19. The results, calculated from Figs. 12, 15, and 19, are shown. (From Keyes [4].)

The forecasts of integration level, pt product, and cooling capability (Figs. 15, 17, and 18) can be used in a physical theory to predict the progress of packaged logic delay. The limiting delay is regarded as the sum of two parts, the circuit delay and the chip-to-chip transit time [35,36]. The latter, however, increases with power; since the power density is limited to values shown in Fig. 17, the more power expended per chip, the further apart the chips must be placed. The power per circuit can be chosen to minimize the total delay.

The use of a power–delay product to characterize technology, as in Fig. 3, implies that device design changes within a given technology can be used to trade speed with power while maintaining a fixed product, i.e., if p is the power per circuit and t_c the delay per circuit, then the semiconductor technology only requires that

$$pt_c = U. \tag{10}$$

The extrapolation of the pt product in simple circuits was shown as a function of time in Fig. 18. In complex logic circuits it depends on circuit fan out and wire loading. Energy must be supplied to change the capacitance of several other devices and of the wiring that interconnects devices in actual logic circuits. A term proportional to the average wire length is added to the pt product of Fig. 3 to take account of the effect of the wiring. The term proportional to length has the form $C'V^2l$, where C' is the capacitance per unit length, V a voltage, and l the total wire length driven by the circuit, taken to be three times the average wire length given by Fig. 14. The value of $C'V^2$ is about 2 pJ/cm in advanced modern chips; a value 1 pJ/cm is used here to allow room for improvement in wire geometry and reduction of voltage. In Summary, the energy per operation used in the model is

$$U = pt + 3 \times (1 \text{ pJ/cm}) \times l_{av} \times (A/N)^{1/2}, \qquad (11)$$

where pt must be understood to mean the power–delay product given by Fig. 3 and A/N has the meaning described in Section V.A.

An analytic representation of l_{av} is used in the calculations. From Fig. 14,

$$l_{av} = 0.3N^g, \qquad (12)$$

where g has the value 0.325.

Delays encountered by signals in passing from chip to chip also constitute an appreciable part of the average logic delay. The contribution of the off-chip delays to the average depends on the frequency with which signals leave a chip and the length of the off-chip wiring on the package. Little accurate information that sheds light on either of these quantities is available. Almost every logic stage sends its output off chip in systems with low levels of integration, and it appears that the off-chip delays amount to the time needed for a signal to traverse about ten chip pitches, where chip pitch is the center-to-center distance between chips on a planar package. This component of delay might be expected to decrease with increasing levels of integration. More lengthy sequences of operations can be completed on a chip if it contains many logic gates, and a signal leaving a chip would seem to have a greater probability of having a receptor at a shorter distance if the level of integration were high. These results are obvious in the limit in which a system can be constructed from only one or a very few chips. Thus, the weighting of off-chip delays used here is essentially an interpolation between the simple views of very low and very high levels of integration just described. The form of the assumed relationships is plotted in Fig. 21, where the average number of chip pitches included in the logic delay per stage is plotted as a function of

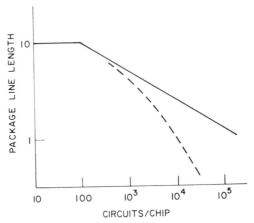

Fig. 21. The weighting factor m for off-chip delays in the minimization of average packaged delays. The solid line is an interpolation between values $m = 10$ for low levels of integration and $m = 1$ for integration approaching the total system on a single chip. The dashed line shows the additional reduction in m that results from the reduced incidence of off-chip connections as the level of integration increases. This factor is based on the assumption that off-chip delays are only encountered with probability $\frac{1}{12}$ if the number of circuits per chip exceeds 50,000. (From Keyes [4].)

the number of logic circuits per chip. These numbers are the values of the quantity m in the theory of packaged delay given below.

Let semiconductor chips, each containing N logic gates with the property described by Eq. (10), be packaged on a planar substrate from which heat can be removed at a maximum density Q per unit area, with each chip occupying an area L^2. Assume further that an amount of power P_D is dissipated in the high-current circuits that drive the transmission lines connecting one chip to another. Then heat removal requires

$$P_D + Np \le QL^2. \tag{13}$$

The time taken for a signal to propagate from one chip to another is mL/c_1, where c_1 is the velocity of electromagnetic waves on the interconnections and m the length of interconnections measured in units of L, the distance between chips. It is also required that L be larger than the chip edge a. The total packaged delay t_p is the sum of the circuit delay t_c and the chip-to-chip propagation time. All of these effects are taken into account by expressing t_p as a function of the power as

$$t_p = (U/p) + (m/c_1)\{[(Np + P_D)/Q]^{1/2} + \alpha\}, \tag{14}$$

where m is the factor that weights the chip-to-chip delays shown in Fig. 21. The value of p that minimizes t_p is given by the equation

$$(1 + Np/P_D)^{1/2} = mNp^2/2c_1UQ^{1/2}P_D^{1/2}. \tag{15}$$

If $P_D = 0$, t_p is a minimum for [36]

$$p = p_0 = (4U^2c_1^2Q/m^2N)^{1/3}. \tag{16}$$

Then

$$t_p = t_0 = (27/4)^{1/3}N^{1/3}(m^2/c_1^2Q)^{1/3}. \tag{17}$$

The effect of $P_D > 0$ is to increase the values given by Eqs. (16) and (17) by factors shown in Fig. 22. Here the ratios p/p_0 and t_p/t_0 are plotted as functions of P_D/Np_0.

The results given by Eqs. (16), (17) and Fig. 22 are used to determine the packaged delays. The resulting logic delay is shown in Fig. 23. The calculation predicts that it will decrease through more than another order of magnitude through the remaining part of the century.

For comparison, an estimated main memory access time is also shown in Fig. 23. A simple view of memory access time was used: It was assumed that the power–chip access time product was proportional to the

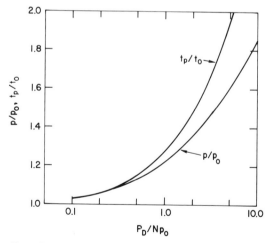

Fig. 22. The effect of driver power on the delay and power obtained from Eq. (15), normalized in the way described by Eqs. (16) and (17).

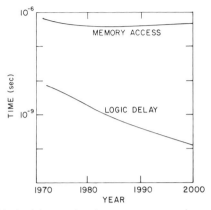

Fig. 23. Packaged logic delays and main memory access times calculated from the extrapolations of technology. (From Keyes [4].)

length of lines on the chip, specifically to the square root of the chip area and that the memory access time increased slowly with the number of chips. Thus [4],

$$T_\text{M} = KC^h p^{-1} A^{1/2}, \qquad (18)$$

where K is a constant, C the number of chips in the memory, p the chip power, A the chip area, and h a small exponent taken to be 0.2. The memory access time does not decrease significantly through the time period in question because of the rapid growth in memory size. It has been assumed that the main memory will continue its historical trend to grow in proportion to computer power, resulting in the number of bits per system shown in Fig. 24.

The increasing gap between logic speed and memory access will have to be taken up by increasing attention to fast buffer or cache memory. The instruction time taking account of the effect of the buffer memory can be expressed in the form [37]

$$T_\text{I} = T_0 + rT_\text{M},$$

where r is the miss ratio, i.e., the probability that a sought-for memory word will not be found in the buffer and access to the main memory will be required, and T_0 is the instruction time if r could be made to vanish. Thus r is reduced as the size of the buffer memory is increased. However, the fast buffer memory is likely to be expensive, so a moderate buffer size

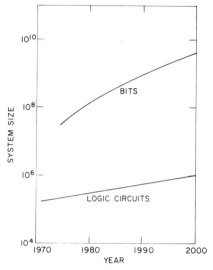

Fig. 24. The rapid growth of memory utilization resulting from the assumption that the number of bits grows in proportion to MIPs. (From Keyes [4].)

tends to optimize the cost of executing instructions. A model that relates the cost of buffer memory chips to the cost of logic chips can be used to find the optimum size [4]. The final result forecasts the growth of large system power, as measured by millions of instructions per second, and is shown in Fig. 25.

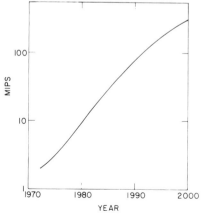

Fig. 25. The projected growth of processing power as measured by millions of instructions performed per second. (From Keyes [4].)

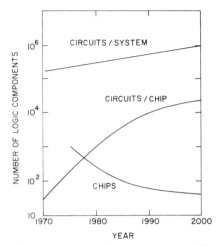

Fig. 26. A summary of the projected evolution of numbers of logic components. (From Keyes [4].)

B. Physical Characteristics of the VLSI System

A few additional results follow from the calculations presented:

(1) The system parameters, that is, total component count, components per chip, and number of chips, are summarized in Figs. 26 and 27 for logic and memory, respectively.

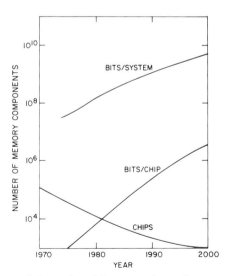

Fig. 27. A summary of the projected future numbers of memory components. (From Keyes [4].)

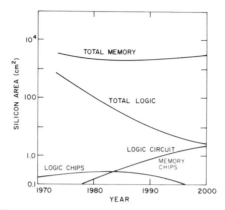

Fig. 28. The projected utilization of silicon area. (From Keyes [4].)

(2)   The assumptions about lithography and level of integration lead to values for chip area. The projections relating to area are summarized in Fig. 28. The memory chip grows steadily in area through the period. The rapid growth of total memory compensates the advantages gained from lithography and ingenuity to keep the total silicon area in the memory nearly constant. On the other hand, the limitation of the size of logic chips to 25,000 circuits implies a decreasing chip size, and the slow growth of circuit count in the system leads to a decrease in the silicon area required for logic. Thus, over 100 times more silicon area will be devoted to memory than to logic by the end of the period.

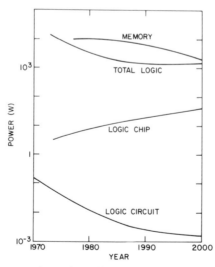

Fig. 29. The power demands predicted by the model. (From Keyes [4].)

(3)   Power utilization by the electronic components is summarized in Fig. 29. Power consumption is increasingly dominated by memory during the period.

VII. ALTERNATIVES

A. Low Temperatures

The problems and limitations of silicon electronics have led to a search for new directions to relax the limitations and solve the problems described. Several possibilities have received attention. One of these involves operating silicon devices at low temperatures, typically 77 K, the boiling temperature of liquid nitrogen [10]. Physics immediately suggests many possible advantages to operating silicon devices at low temperatures. Most obvious is the increase in conductivity; the scattering of electrons by lattice vibrations decreases as the temperature is lowered. The effect of the decreased resistance is probably most important in the metallic interconnections of integrated circuits, since conductivity in the silicon itself is primarily limited by scattering by impurities rather than lattice vibrations. Low temperatures also promise to reduce the power dissipated. As previously explained, the sharp transitions between states of a device that are the essence of digital electronics require that voltages that are large compared to the thermal voltage kT/q must be used. Thus, the lower the temperature, the lower the voltage required in switching circuits. Figure 5 shows dramatically how the sharpness with which an FET is turned on by the gate potential increases as temperature is decreased. The power dissipation is expected to be proportional to the square of the voltage [10]. Thus, these rough considerations suggest that 16 times less power will be required at 77 than at 300 K.

Low temperature may lead to even greater decreases in power in memories in which information is stored in charge on a capacitor. The charge is retained by opening a transistor switch; it gradually leaks away as reverse current through some $p-n$ junction and must be periodically refreshed. The reverse current of junctions decreases rapidly with decreasing temperature, so refreshing is needed less frequently at low temperatures. In some circuits the power dissipated in the refreshing operations is an appreciable part of the total dissipation and may be reduced by a large amount [10].

The decreased power dissipation means that less area is required to allow for removal of heat. The lowered resistance of metals allows connecting lines to be made narrower and also decreases areal requirements.

Thus, significantly higher device densities may be achievable at low temperatures. The cost of low-temperature semiconductor electronics should be substantially smaller than that of conventional semiconductor devices. The unanswered economic question is: how large a system is required to justify the added investment in refrigeration by reduction in cost and increase in performance of semiconductor devices?

B. Materials Other Than Silicon

Another new direction in electronics is the development of digital integrated circuits in other semiconductors. In fact, this direction is not really new, as hope that semiconducting compounds of group III and group V elements would replace silicon and germanium in transistor applications has been entertained ever since the discovery of the compounds more than 20 years ago [38]. A large share of the enthusiasm for the III–V semiconductors was based on their high electron mobilities, which are in many cases from 2 to 20 times or more greater than mobilities in silicon. The search for physical limits to miniaturization reveals one reason why the III–V semiconductors have failed to displace silicon; the limits have little to do with mobility but concern such things as avalanche breakdown fields and thermal problems. It is necessary to have an energy gap that is large enough to prevent intrinsic conductivity, i.e., the thermal excitation of electrons from the valence band to the conduction band, from interfering with device operation. The favorable features of a semiconductor are a high energy gap, which allows the temperature of a component to rise a certain amount and decreases the stringency of the cooling requirements, and high breakdown fields, which are associated with a large gap. A semiconductor with an energy gap smaller than that of silicon is unlikely to play an important role in modern electronics [39].

An even more important reason for the dominance of silicon, however, was the rapid establishment of a feasible processing technology for silicon. The ease with which an SiO_2 layer can be formed on a silicon surface and the remarkable properties of such a layer as an insulator, a diffusion mask, and a neutralizer of undesirable surface effects are unmatched by any phenomena in the III–V compounds.

Nevertheless, the III–V compounds have had an impact on electronics. The advantages of gallium arsenide as a material for transistors have long been known. It has a larger energy gap than silicon, slightly larger breakdown fields, and a much higher electron mobility. The more difficult technology of GaAs has at last been mastered, and it has become the superior material for microwave transistors [40]. Field-effect transistors are used to take advantage of the high electron mobility; in bipolar transistors both electron and hole mobilities are important. One other semiconductor, indium phosphide, appears to share the advantages of GaAs, but its technology is still in a more primitive stage.

The technology that makes GaAs microwave transistors possible is now being extended to the fabrication of integrated microwave circuits. In the future the application of the proved high-speed, high-frequency capabilities of GaAs to digital circuits will begin; exploratory attempts have already been made [11]. A realistic view suggests, however, that the invasion of large-scale digital applications by GaAs will be much more difficult than the conquest of the microwave field. The reason is that one or a few microwave devices with superior frequency response can extend the bandwidth of a system and have great economic value. On the other hand, digital systems use thousands to millions of devices, and low-cost fabrication and high reproducibility are essential. Fast devices are less important because the speed of a system is also limited by the delays in the package. The highly developed and versatile silicon technology, optimized by many years of experience, will not be displaced easily.

C. Superconducting Electronics

Another line of attack on the problems and limitations of silicon electronics is based on the use of superconducting digital devices [41–43] with use being made of nonlinear phenomena associated with the Josephson effect.

A quarter century of experience with semiconductor devices has produced a large body of knowledge concerning the requirements and problems of solid-state electronics. Furthermore, the technology for fabricating superconducting devices and circuits is similar to that used to fabricate semiconductor components, involving such things as deposition of thin films, growth of oxide layers, and photo- or electron-beam lithography. Thus semiconductor electronics is a natural starting point for discussing superconducting electronics and a comparison of the two provides a useful framework for bringing out the reasons for interest in superconducting devices. There are a considerable number of points of distinction between room-temperature, semiconductor electronics and cryogenic, superconducting electronics. It will be most instructive to consider them one by one.

1. Voltage

The voltage used in switching electronics must be greater than kT/q, where q is the charge on an electron. (See Section II.) Thus, cryogenic circuits can operate with smaller voltages than 300 K circuits; ordinary semiconductor circuits require supply voltages of 1 V or more, while superconducting electronics operates from voltage sources of 0.01 V or less. The lower voltages generally imply greater freedom from electric breakdown phenomena. Low voltage also leads directly to low power.

2. Power

The power dissipation of a 4 K superconducting logic circuits is 100 to 1000 times smaller than that of conventional, high-performance, silicon electronics. Since the size of circuits is determined by contemporary lithographic capabilities, this means that the power density is 100 to 1000 times lower in superconducting electronic chips and modules. The abilities of cooling technology to remove heat from a surface do not differ by nearly as large a factor. Therefore, superconducting logic chips can be packaged in much closer proximity to one another in the construction of a large system and the time needed for electrical signals to travel from one part of the system to another is reduced.

3. Speed

The response of semiconductor devices is slowed by the finite time required for electrons to travel through a structure, say the base of a bipolar transistor, and by capacitive effects associated with stored nonequilibrium concentrations of charge carriers. Superconductors, operating through tunneling phenomena, do not show such effects. It seems unlikely, however, that raw device speed considerations will be the important factor in determining the performance of large systems. Rather, the ability to pack substrates very closely together, which is limited by power dissipation and heat removal as previously discussed, will dominate.

4. Resistance

As the dimensions of integrated circuits are reduced, the cross-sectional area of interconnections decreases and their resistance increases in conventional semiconductor circuitry. This increase in interconnection resistance with the progress of miniaturization is one limit to the exploitation of advanced lithography in integrated circuit technology. The vanishing resistance of superconducting interconnections avoids this limit in superconducting electronics.

The vanishing resistance of superconductors also makes persistent currents possible, i.e., currents that can circulate in a loop for many years with no dissipation and no source of power. Persistent currents can be used to store information in nonvolatile form. They can also be a source of difficulty if unintentionally induced in a body of superconducting material.

5. Miniaturization

The miniaturization of semiconductor devices is limited by the existence of depletion regions that support the potential difference between n- and p-type regions of a semiconductor and are an essential part of a semiconductor device. These regions are depleted of mobile carriers and the potential variation within them is controlled by Poisson's law. The thickness of the depletion layers can be reduced by increasing doping levels, but the extent to which this direction can be pursued is limited by the decrease in dielectric breakdown voltage with increasing doping. (See Section IV.B.)

A rather different kind of boundary effect is important in superconducting devices and conductors. Here the Meissner effect, the exclusion of magnetic flux from the interior of a superconductor, is produced by the screening of flux by a current flowing in a thin layer on the surface. The thickness of this layer is known as the penetration depth. The penetration depth is not a limit to the size of superconductors, which can be much smaller than this dimension. A large inductance is associated with such small dimensions, however. The kinetic energy of the electrons that carry the supercurrents is proportional to the square of their velocity and to the square of the current and, consequently, appears as an inductance in a circuit. If the dimensions of a conductor are decreased below the penetration depth, current density, electron velocity, and the inductance increase. If this nongeometrical or kinetic inductance becomes appreciable, circuit performance is degraded [44]. Thus some dimension in the neighborhood of the penetration depth constitutes a practical limit to the size of high-performance, electronic device structures.

6. Alloy Characteristics

As is well known, the electrical properties of semiconductors are controlled by doping with donors and acceptors. The concentrations of dopants employed range from 1% down to 1 part in 10^8. The use of such small concentrations to control properties means that very pure starting materials must be employed and great care must be exercised during processing to prevent contamination. A sensitivity to crystalline defects, which can concentrate impurities or affect their motion in the host material, is created.

Superconductors can frequently be used as undoped materials. Even when alloying elements are used, the concentrations involved are in the range of a few percent. Sensitivity to defects and contaminants is far less than is the case in semiconductors. One aspect of device technology is thereby simplified.

7. Lifetime

Bipolar transistors depend on the ability of technology to provide material with a certain minimal rate of recombination of nonequilibrium concentrations of holes and electrons, in other words, a sufficiently long minority carrier lifetime. Recombination is catalyzed by impurities, imperfections, and surfaces, so another source of sensitivity to defects and contamination exists. No comparable effects are known in the superconducting devices used in electronics.

8. Gain

Gain is essential to the operation of logic circuits. Semiconductor devices can provide current gains of 100 or more (a feature which is very useful in circuit design). The superconducting devices described depend on the production of controlled amounts of magnetic flux by combinations of currents for their operation. Gain is sensitive to the geometrical configuration of structures and the controlled production of high gain devices requires tight dimensional control.

9. Quantum Nature

Both semiconducting and superconducting devices utilize the basic quantum of electric charge—the charge on the electron—in their operation. The fact that charge is quantized, so that a body of charge contains a finite number of electrons, offers an opportunity for fluctuations. Specifically, a fluctuation or noise voltage V_N is associated with charge on a capacitor, where V_N is determined by

$$CV_N^2/2 = kT/2. \tag{19}$$

If one interprets V_N^2 as a fluctuating number of electrons through the relation $n_e = CV_N/q$, then

$$n_e^2 = CkT/q^2. \tag{20}$$

Another quantal effect is also encountered in superconductivity: the quantization of magnetic flux, given in units of $\phi_0 = 2 \times 10^{-15}$ Wb. A fluctuation energy $kT/2$ is also associated with the storage of energy in an inductor. Thus there is a noise current i_N given by

$$Li_N^2/2 = kT/2, \tag{21}$$

where i_N can be interpreted as a fluctuating number of flux quanta in the inductor through the relation $n_f = Li_N/\phi_0$:

$$n_f^2 = LkT/\phi_0^2. \tag{22}$$

A typical inductance of a device designed to operate at 4.2 K is about 10^{-12} H. Thus, according to Eq. (4), $n_f^2 = 10^{-5}$. The thermal fluctuations of the number of quanta in a device are quite small; it is possible to design devices to operate on single flux quanta.

As is well known, dealing with single electrons is not feasible in electronics. It can be seen by inserting a typical device capacitance of 10^{-13} F into Eq. (2) that $n_e^2 = 10^2$ at 4.2 K and larger at higher temperature. Many electrons must be used in the electronic representation of information.

The basic reason for the difference in the fluctuation magnitudes of the magnetic and electric quanta is that the magnetic quantum is very large in the following sense:

$$\phi_0^2/\mu_0 = 3 \times 10^{-22} \quad \text{J cm}, \tag{23}$$

$$q^2/\epsilon_0 = 3 \times 10^{-25} \quad \text{J cm}. \tag{24}$$

The difference between the magnitude of the quanta is enhanced by material properties and geometrical effects; the values of capacitances are usually enhanced by high dielectric constants and thin dielectric layers.

10. Atomic Motion

The low temperature of a cryogenic system greatly slows the rates of activated atomic motion and chemical reactions. Thus, such causes of degradation and failure of components as electromigration and corrosion will be greatly reduced.

It is possible to see many advantages to a cryogenic superconducting computer. Certain problems, such as that of providing many interconnections between various levels of packaging are not alleviated. The development of the new material systems required will be time-consuming and expensive. Thus, the role of superconducting electronics in information processing is still not clear.

APPENDIX. MESFETs

For many years it has been possible to predict that the high electron mobility of GaAs should permit very fast transistors to be made in this material [44]. The potential of GaAs has only been realized through the MESFET, however. One-micron MESFETs were constructed in silicon and in GaAs in the late 1960s and immediately they greatly extended the useful frequency range of transistors [45–47]. The MESFET uses the high electron mobility of GaAs to advantage, so that GaAs is almost exclusively used in preference to silicon in very high frequency microwave FETs [40]. The operation of MESFETs in simple digital circuits has also been demonstrated, with many possibilities for the future of MESFETs in digital devices to be considered [11,12,48,49].

The basic idea of the MESFET is shown in Fig. 6. The device consists simply of a thin layer of conductive semiconductor on a high resistivity substrate with a Schottky diode gate between two metal source and drain contacts. A depletion region is associated with the Schottky barrier. The thickness of the depletion region depends on the voltage applied to the Schottky diode; it can be made thick enough to deplete the thin conductive layer entirely. Thus, voltage applied to the gate can control the flow of current between the source and the drain. The use of a depletion layer to control the conductance between two electrodes is reminiscent of the junction FET, and the theory of the MESFET and the JFET are essentially identical.

A few specific points about MESFETs should be noted: There has been less incentive to pursue MESFETs in silicon than in GaAs because of the availability of SiO_2 as a gate insulator. Usually MESFETs are operated in a depletion mode, where a reverse bias is applied to the gate to deplete the thin conductive layer of mobile carriers. Enhancement mode or normally off MESFETs can however, be made; it is simply required that the conductive layer be made thinner than the depleted layer that arises from the built-in metal–semiconductor barrier in the absence of applied potentials.

A. MESFETs versus IGFETs

The outstanding difference between the MESFET and the IGFET is that the former does not require the deposition of a thin insulating layer. This difference is crucial in the case of GaAs, where the technology for providing the needed high-quality insulator is not known. Even in silicon, however, a simpler technology would be advantageous in perhaps reducing defect levels and thus making larger chips possible. The simplicity of MESFETs is revealed in the published literature: 1-μm gate length MESFETs are common; 1-μm IGFETs are unusual. In fact, 0.5-μm MESFETs were realized in 1973 [50].

Elimination of the SiO_2 would also remove the undesirable trapping effects that are associated with it. A pursuit of simpler technology also leads one to the use of normally off MESFETs, which use a simpler circuit than depletion mode devices. The normally off mode of construction has other subsidiary advantageous features: Since Schottky barriers completely deplete the conductive layer, they can be used to form isolating boundaries in the gate formation process step and the thin layers have a high sheet resistance, which is useful for fabrication of resistors. Voltages and power dissiplation are less than in depletion-mode devices.

Two disadvantages of normally off MESFETs appear when they are compared with IGFETs. First, a forward bias must be applied to the Schottky diode gate to turn the device on. Gate current flows under the influence of the forward bias. It must turn out that the voltage that induces adequate channel conductance does not draw an intolerable amount of gate current. The operating voltage may be confined within rather narrow limits. This feature can be turned to advantage, however, by using the diode action as a clamp for the upper logic level in normally off circuits. Maintaining a uniform operating range over a large number of devices requires careful control of the thickness and concentration of the doped layer. Ion implantation appears to be evolving as the preferred method of preparing the conducting layer.

Second, present IGFETs provide better utilization of semiconductor area than the MESFET. This point is illustrated in Fig. 6, in which it can be seen that a MESFET requires an appreciable separation between the source and drain electrodes and the gate. It also follows from the structures shown in Fig. 6 that the gate–drain capacitance is small in a MESFET but that a series resistance is introduced into the channel. Indeed, most MESFETs will have more complicated structures than shown in Fig. 6 because heavily doped regions will be created under the source and drain contacts and extending toward the gate to reduce the series resistance.

B. Extendability

The MESFET is an extremely simple device. Current MESFETs appear to be truly limited by lithography. The depletion layers in current MESFETs, which use doping levels of about 10^{17} cm^{-3}, are only 1000 Å wide. There is no reason other than lithographic capability that the gate lengths of more heavily doped MESFETs could not be 1000 Å and the source/drain–gate spacings in a normally off device could be only slightly greater, say 1500 Å. In other words, substantial miniaturization is possible with the presently used vertical structure. The effect of short gates on silicon MESFETs has been investigated by simulation [51]. The simulation shows that although transistor characteristics are modified by short-channel effects—for example, the pinch-off voltage is altered and the output conductance increased—devices are workable and their speed increases with decreasing gate length as long as the length is greater than the thickness of the conductive layer. Thus, the MESFET is seen to be an ideal vehicle for the utilization of advanced lithographic techniques and attainment of high device densities.

C. Silicon versus GaAs

The principal advantage of silicon in developing a digital MESFET technology would be the ability to draw on the great store of well-developed techniques already available for fabricating structures in silicon. In addition, knowledge acquired by MESFET development in silicon would have a certain applicability to other uses of Schottky barriers in silicon technology. The widely established use of silicon provides a ready market for new developments in silicon devices. Against this, it must be observed that almost all of the physical properties of GaAs are more favorable to devices than those of silicon.

REFERENCES

1. G. E. Moore, *in* "Solid State Devices 1977" (E. A. Ash, ed.) Conference Series Number 40, pp. 1–6. Institute of Physics, London.
2. R. W. Keyes, *Science* **196**, 945–949 (1977).
2a. R. W. Keyes, *Science* **195**, 1230–1235 (1977).
3. G. Marr, Perspectives on MOS Directions, "Fall Compcon 77 Digest" (Catalog No. 77CH1258-3C), pp. 242–244. IEEE, New York.
4. R. W. Keyes, *IEEE Trans. Electron Devices* **ED-26**, 271–279, (1979).
5. K. U. Stein, *Electrotech. Maschinenbau* **93**, 240–248 (1976).
6. R. W. Keyes, *Proc. IEEE* **63**, 740–767 (1975).
7. C. S. Shannon, *Bell Syst. Tech. J.* **27**, 379–423 (1948).
8. J. von Neumann, "The Computer and the Brain." Yale Univ. Press, New Haven, Connecticut, 1958.
9. A. W. Lo, *IRE Trans. Electron. Comput.* **EC-10**, 416–425 (1961).
10. F. H. Gaensslen, V. L. Rideout, and E. J. Walker, *Tech. Digest Int. Electron Devices Meeting*, pp. 43–46. IEEE, New York, 1975; F. H. Gaensslen, *et al.*, *IEEE Trans. Electron Devices* **ED-24**, 218–229 (1977).
11. R. C. Eden, B. M. Welch, R. Zucca, and S. I. Long, *IEEE Trans. Electron Devices* **ED-26**, 299–317 (1979).
12. G. Nuzillat, C. Arnodo, and J. P. Puron, *IEEE J. Solid State Circuits* **SC-11**, 385–394 (1976).
13. R. H. Dennard, F. H. Gaensslen, E. J. Walker, and P. W. Cook, *IEEE Trans. Electron Devices* **ED-26**, 325–333 (1979).
14. G. Pircher, *in* "Solid State Devices, 1975," pp. 31–72. Societe Francaise de Physique, Paris, 1975.
15. J. T. Wallmark, *in* "Solid State Devices, 1974," pp. 133–167. Institute of Physics, London, 1975.
16. G. R. Brewer, *IEEE Spectrum* **8** (1), 23–37 (1971).
17. A. V. Crewe, J. Wall, and J. Langmore, *Science* **168**, 1338 (1970).
18. T. E. Everhart, unpublished calculations (1976).

230   R. W. Keyes

19. N. W. Parker, S. D. Golladay, and A. V. Crewe, *Scanning Electron Microscopy, 1976* 37–44.
20. Y. Tarui, "Technical Digest 1977 IEDM," pp. 2–6. IEEE, New York, 1977.
21. E. D. Wolf, *Phys. Today*, pp. 34–36 (November 1979).
22. H. I. Smith, and S. E. Bernacki, *J. Vac. Sci. Technol.* **12**, 1321–1323 (1976); R. Feder, E. Spiller, and J. Topalian, *ibid.* **12**, 1332–1335 (1976).
23. A. N. Broers *et al., Appl. Phys. Lett.* **29**, 596–598 (1976).
24. B. Hoeneisen and C. A. Mead, *Solid-State Electron.* **15**, 819–829, 981–987 (1972).
25. T. H. Ning *et al., IEEE Trans. Electron Devices* **ED-26**, 346–353 (1979).
25a. F. M. Klaassen, *Solid-State Electron.* **21**, 565–571 (1978).
26. W. Shockley, *Solid-State Electron.* **2**, 35–67 (1961).
27. E. H. Nicollian and A. Goetzberger, *Bell Syst. Tech. J.* **46**, 1055–1133 (1967).
28. R. W. Keyes, *IEEE J. Solid State Circuits* **SC-10**, 245–247 (1975); *Appl. Phys.* **8**, 251–259 (1975).
29. W. E. Donath, IBM Research Rep. RC4610; W. R. Heller, W. F. Mikhail, and W. E. Donath, *Proc. Design Automat. Conf., 14th,* New Orleans, Louisiana pp. 32–42 (1977); B. S. Landman and R. L. Russo, *IEEE Trans. Comput.* **C-20**, 1469–1479 (1971).
30. E. Bloch and D. Galage, *Computer* pp. 64–76 (April 1978).
31. R. J. Blumberg and S. Brenner, *IEEE J. Solid State Circuits* **SC-14**, 818–822 (1979); W. Braeckelmann *et al., ibid.* **SC-14**, 829–832 (1979).
32. For a detailed description of the packaging of a modern computer see R. J. Beall, in "1974 INTERCON Technical Papers," paper 18/3. IEEE, New York, 1974.
33. E. A. Wilson, *1977 Nat. Comput. Conf. Proc.* pp. 341–348. AFIPS Press, Montvale, New Jersey, 1977.
34. R. M. Russell, *Commun. ACM* **21**, 63–72 (1978).
35. W. Vilkelis and R. A. Henle, "Spring Compcon 79 Digest" (Cat. No. 79CH1393-BC), pp. 285–289. IEEE, New York.
36. R. W. Keyes, *IEEE J. Solid State Circuits* **SC-13**, 265–266 (1978).
37. J. H. Saltzer, *Commun. ACM* **17**, 181–185 (1974).
38. D. A. Jenny, *Proc. IRE* **46**, 959–968 (1958).
39. C. A. Liechti, *IEEE Trans. Microwave Theory Tech.* **MTT-24**, 279–300 (1976).
40. R. W. Keyes, *Comments Solid State Phys.* **8**, 37–53 (1977).
41. H. H. Zappe, *IEEE Trans. Mag.* **MAG-13**, 41–47 (1977); "Solid State Devices 1976" (R. Mueller and E. Lange, eds.), pp. 39–54. Institute of Physics, London, 1977. pp. 39–54.
42. K. K. Likharev, *IEEE Trans. Magn.* **MAG-13**, 242–244 (1977).
43. T. A. Fulton and L. N. Dunkelberger, *Appl. Phys. Lett.* **22**, 232–233 (1973); T. A. Fulton, R. C. Dynes, and P. W. Anderson, *Proc. IEEE* **61**, 28–35 (1973).
44. M. Klein, Josephson tunneling logic gates with thin electrodes, *IEEE Trans. Magn.* **MAG-13**, 59–62 (1977).
45. K. E. Drangeid *et al.,* Microwave silicon Schottky-barrier field-effect transistor, *Electron Lett.* **4**, 362–363 (1968).
46. K. Drangeid, R. Sommerhalder, and W. Walter, *Electron. Lett.* **6**, 228–229 (1970).
47. J. Turner, A. Waller, R. Bennett, and D. Parker, *Symp. GaAs Related Compounds, 1970* pp. 234–239. Inst. Phys., Conf. Series No. 9, London, 1971.
48. N. Kawamura *et al.,* IEDM/1970 Abstracts. IEEE, New York, 1970; *Elec. Electron. Abstr.* **74**, 583 (1971).
49. K. Drangeid *et al., IEEE J. Solid State Circuits* **SC-7**, 277–282 (1972).
50. W. Baechtold *et al., Electron. Lett.* **9**, 232–234 (1973).
51. M. Reiser, *Electron. Lett.* **8**, 254–256 (1972).

Chapter **6**

Physics and Modeling of Submicron Insulated-Gate Field-Effect Transistors. I

D. K. FERRY

Department of Electrical Engineering
Colorado State University
Fort Collins, Colorado

I. INTRODUCTION

The concept of space-charge effects at the surface of a semiconductor has been well known for a considerable period of time. Such effects were in fact first studied in connection with metal–semiconductor contacts

231

[1,2], although the suggestion of stable space-charge layers in these systems followed much later [3]. However, this latter result probably contributed to the suggestions at that time for a surface field-effect device [4,5]. Actual use of field-effect devices and the oxide–semiconductor system for these devices followed later [6–9]. Advances in the field were rapid from that point onward, and the metal–oxide–semiconductor field-effect transistor (MOSFET) occupies a major role in large-scale integration (LSI) of electronic circuits, a role that is becoming increasingly dominant as device dimensions continue to decrease.

The metal–oxide–semiconductor (MOS), or metal–insulator–semiconductor (MIS), structure was first proposed as a voltage-variable capacitor [7,8]. The characteristics of the structure were subsequently analyzed by Frank [10] and Lindner [11] and used to study the silicon surface by Terman [12] and Lehovec et al. [13]. Following these initial studies of the properties of the surface, Kahng and Attala [9] fabricated a thermally oxidized, silicon, field-effect transistor. The basic device characteristics have been generally formulated [14–20] and many models developed, as can be seen from recent reviews of these models [21,22].

In the past several years, the integration density of silicon circuits has increased steadily and, when compared to the medium-scale integration of a decade ago, dramatically. A significant part of this steady increase lies in the reduction of channel lengths of the individual devices, and this reduction has been supported by several technological developments such as more accurate process control and fine-pattern lithography by optical, electron-beam, and ion-beam techniques. However, as the channel length is reduced, many effects, which heretofore were of second-order importance, become of primary importance and dominate device and circuit performance. The reduction of device size in order to achieve greater performance has followed a scaling principle, but this approach is limited by physical and practical problems. An insight into some of these problems can often be obtained through the study of models of the device itself.

The traditional approach to the modeling problem of semiconductor device operation is to divide the device into several areas for which several different, but linear, approximations hold. Then, the one-dimensional transport problem is solved consistently within these areas, such as with the gradual-channel approximation. However, it is now well recognized that such approximate approaches are often inappropriate for submicron devices. In order to gain the complete insight and understanding of device behavior for the submicron device, more general and accurate two-dimensional (at a minimum) solutions are required and these are generally only obtainable by numerical techniques [23–33]. Suitable computer codes for these solutions are thus becoming an important part of the device designers' repetoire. The solution to the two-dimensional Poisson's equation represents no conceptual difficulties and the major physical effects are dominantly tied up in the manner in which the charge fluctuations and current response are coupled to the local electric field, formally related through the continuity equation. However, these detailed computations often give only detailed results that verify our more direct intuitive ideas on the principal physics of operation of the device. Between our intuition and these detailed calculations, a rather good qualitative understanding of the physics germane to submicron devices has been obtained, and it is this qualitative understanding that we wish to review in this chapter.

In the following section, we shall treat the MOSFET itself, covering not only the gradual-channel approximation and saturation, but also physical limits and scaling. Then, in Section III, effects that arise from the two-dimensional nature of the device and the necessary charge sharing in small devices will be considered. There, we will also treat subthreshold currents, drain-induced barrier lowering, and the drain dependence of saturation. Hot-carrier injection and avalanching effects will be treated in Section IV, and, finally, carrier transport will be discussed in Section V. The entire discussion will be limited to the Si–SiO$_2$ system, due to its almost universal usage in LSI. While an attempt has been made to include significant references, no attempt was made to include an exhaustive survey of the literature.

II. THE MOSFET

A typical n-channel MOSFET structure is shown in Fig. 1. Two n-type regions are diffused into the p-type substrate. These regions form the drain and source contacts. The gate structure is essentially combined with the p-type substrate to form an MOS diode. If the gate is biased positively, a negative surface space charge appears at the semiconductor surface next to the oxide. For a sufficiently large forward bias, an n-type inversion layer forms at the surface. This inversion layer forms a narrow channel between the source and the drain contacts, and this inversion channel conducts current from the source to the drain. The MOS structure modulates this current by varying the surface charge. The depth of the channel into the p-region is determined by the gate voltage and the drain voltage, since it is the difference in voltage across the region, $V_G - V(x)$, where $V(x)$ is the surface potential at x, which determines the

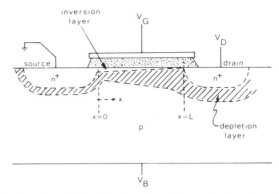

Fig. 1. The insulated-gate field-effect transistor. The conduction channel is formed by a surface inversion layer under the MOS capacitor structure.

surface-charge density at that point. From this, it is apparent that the voltage across the oxide layer and channel is a decreasing function of x, as x increases from the source to the drain. Hence, the channel width decreases slightly as we move from the source to the drain. Pinch-off occurs first at the drain end of the channel, and it occurs when the voltage is no longer able to maintain the inversion layer at that point. At this value of drain voltage, the drain current saturates.

The basic mode of operation, the enhancement mode, utilizes the approach just discussed. It is possible, however, that the work-function difference and oxide charge, or even an implanted n-layer in the channel region, can lead to a surface channel existing when no gate voltage is applied. In this case, the device is termed a depletion-mode device since gate voltage is used to turn the device off rather than on. The depletion-mode device is often used as an active load device in integrated circuits and is traditionally fabricated by ion-implantation techniques. Several recent models of such a device have appeared [34–41]. Other variations of the basic MOS transistor that have appeared have placed the channel on a groove (V-groove of VMOS) (see, for example, [42]), have used diffused regions near the drain for higher voltage operation (DMOS) (see, for example, [43–45]), or have left the gate floating for memory applications (FAMOS) (see, for example, [46,47]).

A. The Gradual-Channel Approximation

In order to model the MOS transistor in a relatively exact manner, it is necessary to account for the charge variation along the channel by writing differential equations. The incremental voltage drop along the channel is represented as a function of the current through the differential impedance. Integration of this equation leads to a relationship for the drain current in terms of the applied voltages. We assume that the gradual-channel approximation is valid; that is, the fields in the direction of current flow are much smaller than the fields normal to the silicon surface. This assumption validates the use of a one-dimensional MOS analysis to find the carrier concentrations and the dimensions of the depletion region under the gate. This is equivalent to ignoring that portion of Poisson's equation arising from $\partial\Psi/\partial x$ (along the channel), allowing a one-dimensional Poisson's equation to be retained. In practice, the fields normal to the surface are generally at least an order of magnitude larger than fields along the channel, except in the saturation region. For the present, we also assume that the channel length L is large compared to the depletion widths at the source and drain junctions.

Strong inversion at the surface will occur when the minority carriers at the surface become equal to the majority carrier density in the bulk or, for p-type material, $n_s = p_0$. This occurs when

$$\Psi_s = -2\phi_b = 2 \ln(N_a/n_i), \tag{1}$$

where N_a is the acceptor concentration in the p-type substrate, n_i the intrinsic concentration, the reduced bulk potential is

$$\phi_b = (E_F - E_{Fi})/k_B T,$$

and

$$\Psi_s = \phi_s - \phi_b = eV_s/k_B T$$

is the reduced surface potential. Thus the critical turn-on voltage for the channel is

$$V_T = k_B T \Psi_s/e = -2(E_F - E_{Fi})/e = (2k_B T/e) \ln(N_a/n_i). \tag{2}$$

Actually, we shall include the possible effects of difference in the work functions and surface states through the flat-band voltage V_{FB}, and this will modify the turn-on voltage somewhat. Pinch-off occurs when V_D increases to a value such that

$$V_G - V_D = V_T$$

at the drain end of the channel. If we have a silicon substrate with $N_a = 10^{16}\ \mathrm{cm^{-3}}$, we find that $V_T - V_{FB} = 0.72$ V, so that this would be the turn-on voltage in the absence of any surface states or work function differences.

We follow a procedure in which the current flow through the channel is calculated as a function of drain voltage. For FET operation, we assume that $V_G \gg V_T$. Then, an increment of resistance along the channel is

$$dR = dx/Z\mu_e\rho_{s,n}(x),$$

where $\rho_{s,n}(x)$ is the surface-charge density in the inverted layer, Z the lateral extent of the channel, and μ_e an effective electron mobility. The total surface charge is

$$\rho_{s,n} = \rho_s - eN_aW, \tag{3}$$

where W is the width of the depleted region. At the onset of inversion, W becomes essentially constant. The applied gate voltage V_G divides between the capacitance of the oxide layer and the surface potential V_s. This is given by

$$V_G - V_{FB} = +(\rho_s/C_0) + V_s, \tag{4}$$

where we have added V_{FB}, the flat-band voltage, to take account of surface states and work function differences. We can combine Eqs. (3) and (4) to give $\rho_{s,n}$ as

$$\rho_{s,n} = +C_0(V_G - V_{FB} - V_s) - eN_aW. \tag{5}$$

The width is just

$$W = [(2\epsilon/N_ae)(V_C - V_B - 2\phi_bk_BT/e)]^{1/2}, \tag{6}$$

where V_C is the voltage $V(x)$ in the channel and V_B the substrate bias. For simplicity, we write

$$V_{TB} = -(2\phi_bk_BT/e) - V_B \tag{7}$$

as the bulk and substrate contributions to the turn-on voltage. In the following, we also replace V_C with $V(x)$, and

$$W = \{(2\epsilon/N_ae)[V(x) + V_{TB}]\}^{1/2}. \tag{8}$$

Combining Eqs. (5)–(8) gives

$$\rho_{s,n} = +C_0[V_G - V_T' - V(x)] - \{2\epsilon N_ae[V(x) + V_{TB}]\}^{1/2}, \tag{9}$$

where

$$V_T' = V_T + V_{FB}. \tag{10}$$

The increment of channel resistance is then just

$$dR = \{C_0[V_G - V_T' - V(x)] - (2\epsilon N_ae)^{1/2}[V(x) + V_{TB}]^{1/2}\}^{-1} dx/Z\mu_e, \tag{11}$$

where Z is the channel width. In writing $V(x)$ as we have, $V(x) = 0$ defines the source end of the channel.

The voltage drop along the increment of channel arises from the resistance drop in the increment of length dx as

$$dV = I_D \, dR$$

and we can integrate this over the channel length. Thus

$$Z\mu_e C_0 V_D[V_G - V_T' - \tfrac{1}{2}V_D]$$
$$- \tfrac{2}{3}Z\mu_e(2\epsilon N_a e)^{1/2}[(V_D + V_{TB})^{3/2} - V_{TB}{}^{3/2}] = I_D L$$

and

$$I_D = (Z\mu_e C_0/L)[V_G - V_T' - \tfrac{1}{2}V_D - (\bar{Q}_B/C_0)]V_D, \qquad (12)$$

where

$$\bar{Q}_B = (1/V_D)\int Q_b(x)\,dV = \tfrac{2}{3}(2\epsilon N_a d)^{1/2}[(V_D + V_{TB})^{3/2} - V_{TB}{}^{3/2}]/V_E \qquad (13)$$

is an average of the depletion charge along the channel and Eq. (12) is valid for $V_D < V_P$, the pinch-off voltage. For $V_D \ll V_{TB}$, we can approximate the expression for $\bar{Q}_B$ by expanding the terms in brackets and

$$\bar{Q}_B = \rho_{s,0} = (2\epsilon N_a e V_{TB})^{1/2}$$

and

$$I_D = (Z\mu_e C_0/L)[V_G - V_T' - \tfrac{1}{2}V_D - (\rho_{s,0}/C_0)]V_D.$$

When the drain voltage is increased to a level that the mobile charge in the inversion layer $\rho_{s,n}(L) = 0$, pinch-off has occurred at the drain. At this point, the current saturates at I_{Dsat} and occurs for $V_D = V_{Dsat}$. The value of V_{Dsat} is obtained from the condition $\rho_{s,n}(L) = 0$ or from Eq. (9) with

$$C_0(V_G - V_T' - V_{Dsat}) = [2\epsilon N_a e(V_{Dsat} + V_{TB})]^{1/2} \qquad (14)$$

or

$$V_{Dsat} = V_G - V_T' + K^2\{1 - [1 + (2V_G/K^2)]^{1/2}\}, \qquad (15)$$

where $K = (\epsilon N_a e)^{1/2}/C_0$ is related to the average depletion charge along the channel. The saturation current is found by using Eq. (15) in Eq. (12). Thus

$$I_{Dsat} = G_0 V_{Dsat}[V_G - V_T' - \tfrac{1}{2}V_{Dsat} - (\bar{Q}_{Bsat}/C_0)], \qquad (16)$$

where $\bar{Q}_{Bsat}$ is $\bar{Q}_B(V_{Dsat})$. For a doping level of 10^{17} cm^{-3} in Si and an oxide thickness of 200 Å, $K \approx 0.25$.

In Eq. (20) for the current, we note that I_D is dependent on the actual channel length L. However, this length is a function of the drain voltage due to depletion-region widening at the drain end of the channel. For increases of V_D above V_{Dsat}, the end of the ohmic channel (the pinch-off point) moves away from the drain. This causes the effective length L', the distance from the source to the end of the channel, to decrease. It is this L' rather than L that should appear in the equations for the saturated current and transconductance. An accurate theory for the effective channel length is complicated by the two-dimensional nature of the potential and the interaction of the gate electrode in this region [48–50]. A simple model can, however, take this into effect, if the channel is not too small [51–55]. For V_D close to V_{Dsat}, the effective channel length can be approximated by a one-dimensional space-charge-spreading analysis of the depletion region around the drain. The distance between the pinch-off point and the drain contact edge is then

$$L - L' = [2\epsilon(V_D - V_{Dsat})/eN_a]^{1/2}, \tag{17}$$

and for $I_D > I_{Dsat}$,

$$I_D = \frac{L}{L'} I_{Dsat} = \frac{\frac{1}{2}Z\mu_e C_0(V_G - V_T'')^2}{L - [2\epsilon(V_D - V_{Dsat})/eN_a]^{1/2}}. \tag{18}$$

B. Scaling the MOSFET

Scaling down the size of the MOSFET in order to achieve increased packing densities means a proportional reduction of device dimensions, pattern geometries, and the power supply voltage. For example, scaling down the size of a device by a factor of 4 can result in a 16 times greater integration density and even greater gain of performance, as indicated by the speed–power product. To achieve this factor, the channel length and width are each reduced by the factor of 4. However, reducing the channel length is limited by the onset of punchthrough, and to avoid this the depletion widths are reduced by reducing the circuit supply voltage and increasing the substrate doping concentration. This creates a further complication, since as the substrate doping is increased, the field required for strong inversion is increased, necessitating a reduced gate oxide thickness. All of these factors taken together lead to a general theory of scaling [56–58].

In general, the length and width of the channel are assumed to be reduced by a factor of α. In addition, the source voltage is also reduced by a factor of α and the sheet resistivity is increased by α (N_a is increased by α). Thus the drain depletion width is reduced by α. The surface electric field at turn-on is approximately eN_aW/ϵ, and since ϕ_b scales only logarithmically with α, the surface field must increase approximately as $\alpha^{1/2}$, which also is a weak dependence. As V_G has decreased by α, the oxide thickness must also decrease by at least α. Usually the oxide thickness is scaled exactly by α, using the fact that the threshold voltage will be reduced in the smaller device, as will be discussed later. Then, the gate capacitance scales as

$$C_{0T}' = \frac{\epsilon_0(A/\alpha^2)}{y_0/\alpha} = \frac{C_{0T}}{\alpha}, \tag{19}$$

but the specific capacitance per unit area (which appears in all of the preceding equations) increases as

$$C_0' = C_0 \alpha. \tag{20}$$

Thus the saturated drain current scales as

$$I_{\text{Dsat}}' = \frac{(Z/\alpha)\mu_e}{2(L/\alpha)} C_0 \alpha \left(\frac{V_G - V_T''}{\alpha}\right)^2 = \frac{I_{\text{Dsat}}}{\alpha} \tag{21}$$

and the transconductance remains unchanged. Thus, the power is reduced by α^2. The transit time through the device is also reduced (assuming a saturated velocity situation) as is the switching speed. The latter scales as C_{0T}/g_m, where g_m is the transconductance, which scales as $1/\alpha$ from Eq. (19). Thus the smaller devices are faster and the speed–power product, or equivalently the energy stored in the gate capacitance, goes as $1/\alpha^3$. But if node capacitance dominates, deviation of the speed–power product away from $1/\alpha^3$ scaling may occur. It is possible actually to generalize these arguments to include the field patterns and current densities, but the field pattern and carrier velocities should be unchanged. Here is where second-order effects such as threshold voltage shift and mobility reduction become significant.

There are limitations to this scaling of device size, however. As previously mentioned, short-channel effects lead to a reduction of the threshold voltage and drain-induced subthreshold currents. These can significantly affect the usage of small devices in actual logic circuit functions. The subthreshold density, for example, is given by

$$\rho_{s,n} \approx -\frac{k_B T}{2} \left(\frac{\epsilon N_a}{2|\phi_b|}\right)^{1/2} \exp\left[\frac{e(V_G - V_T)}{k_B T}\right]. \tag{22}$$

The pre-factor scales as $\alpha^{1/2}$, since ϕ_b varies only weakly with α. The major problem lies in the argument of the exponential, however. The off-state resistance of the MOSFET varies as ($V_G = 0$)

$$R_{\text{off}} \approx 1/\rho_{s,n} \approx \exp(eV_T/k_B T), \tag{23}$$

as this resistance is dominated by the subthreshold effects. Normally, for logic applications, the threshold is set in the vicinity of $V_T \approx 0.2V_{DD}$, where V_{DD} is the drain circuit supply voltage. For $V_{DD} = 5$ V, the argument of the exponential is about 200 and R is large. As V_{DD} is scaled down by α though, the off-state resistance changes as $R^{1/\alpha}$. Thus great care must be exercised to control subthreshold currents in order to maintain an adequate resistance in the off state for logic applications. The problem arises, of course, because the temperature does not scale. It is further complicated by short-channel effects that lead to threshold reductions.

A second limitation on scaling arises from attempts to maintain voltages higher than those required in strict scaling. As fields increase in the device, hot electrons can be emitted from the silicon into the gate oxide. The hot electrons can originate either from the current in the surface channel or from the substrate. Subsequent trapping of the electrons in the oxide can lead to threshold instabilities due to oxide charge effects on the threshold voltage. This will be discussed in more detail in a later section.

A final limitation on the amount of size reduction that can be achieved is from the breakdown voltage of the oxide itself. The breakdown field is about 7×10^6 V/cm in silicon dioxide and this sets a limit on how small devices can be made [59].

C. Channel Quantization

When the bands at the surface are bent strongly, as in strong inversion where the Fermi level approaches the conduction band, the potential well formed by the insulator–semiconductor surface and the electrostatic potential in the semiconductor can be narrow enough that quantum-mechanical effects become important. The motion of the electrons in the direction perpendicular to the surface is constrained to remain within this potential well, and if the thickness is comparable to the electron wavelength, size-effect quantization leads to widely spaced subbands of electron energy levels. The electron energy levels are grouped into these subbands, each of which corresponds to a particular quantized level formation in the direction perpendicular to the surface (for a general review, see [60]). In the case of silicon, the transport of electrons within the inversion layer remains dominated by intervalley and acoustic phonons, at least for temperatures greater than or of the order of liquid nitrogen temperatures, although other scattering mechanisms are also of importance. We shall deal with this transport in a later section.

In a quantized electron inversion layer at the surface of (100)-oriented silicon, the six equivalent minima of the bulk silicon conduction band split into two sets of subbands. One set consists of the subbands arising from the two valleys that show the longitudinal mass in the direction perpendicular to the surface. This set has energy levels $E_0, E_1, E_2, \ldots$ in the notation of Stern and Howard [61]. The lowest subband at the surface E_0 belongs to this set. The other set of subbands arises from the four equivalent valleys that show a transverse mass in the direction normal to the surface. This set has energy levels designated as $E_0', E_1', E_2', \ldots$. Generally, these levels line up such that E_0' is almost degenerate with E_1 of the twofold set. Except at very low temperatures, it is unreasonable to assume that all the electrons are in the lowest subband. However, for most transport calculations, it is not a bad approximation to assume that most of the electrons occupy the three lowest subbands E_0, E_1, and E_0'.

The effect of this quantization is twofold. First, the density of carriers in the inversion layer is generally large enough to generate a self-energy correction to the surface potential, necessitating a self-consistent solution for Ψ including the details of the charge density and its wave functions. Second, the two-dimensional nature of the electrons modifies the density of states and, hence, modifies the transport properties. An estimate of the size of the effect can be found from estimating the subband separation $E_0' - E_0$. In a small device, $N_a - N_d \approx 10^{17}$ cm^{-3}, $\epsilon = 11.8\epsilon_0$, $V_s \approx 1.0$ V, and $W_{max} \approx 0.12$ μm with $N_{depl} = (N_a - N_d)W_{max} \approx 1.2 \times 10^{12}$ cm^{-2}. From a variational calculation, the lowest energy level of each set of subbands can be calculated from (relative to E_c) [61]

$$E_0 = \left(\frac{3}{2}\right)^{5/3} \left(\frac{e^2\hbar}{\epsilon}\right)^{2/3} \frac{N_{depl} + (55/96)N_s}{\{m_3[N_{depl} + (11/32)N_s]\}^{1/3}}. \tag{24}$$

For $N_s = 6 \times 10^{12}$ cm^{-2}, a reasonable level in a device, we find that $E_0' = 280$ meV and $E_0 = 184$ meV, so that $E_0' - E_0 \approx 100$ meV, or roughly $4k_B T$ at room temperature. Thus, we can expect these quantum effects to be pronounced in small, submicron devices.

III. TWO-DIMENSIONAL EFFECTS

Attempts to go beyond the gradual-channel approximation are driven by the need to treat more fully the short-channel effects, such as drain-voltage-induced shifts of threshold, general sensitivity to threshold shifts, and actual channel-mobility variations. Many of these effects fall under the category of second-order effects in that they are germane only for small performance perturbations in large devices. However, in small devices, these effects tend to become very important to device performance and must be considered more closely. In this section, we shall treat the areas of short-channel effects, subthreshold currents, and variations of saturation voltage.

A. Gate–Drain Charge Sharing

It was assumed in the gradual-channel approximation that the induced surface charge was a function only of the voltage differential between the gate and the substrate. This assumption is not valid near the source and drain depletion regions due to the additional fields and charge densities there, thus requiring the full three-dimensional terms in Poisson's equation. If the channel is long, however, these edge regions do not affect the

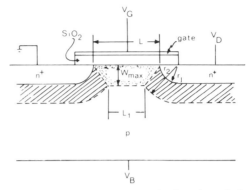

Fig. 2. A short-channel MOSFET, showing effective channel shortening and drain-induced variation of threshold due to reduced depletion charge.

overall treatment of the device by these approximate equations. For short channels though, these edge effects can no longer be ignored. Because of the interpenetrating depletion regions, a significantly smaller amount of charge (in the normal space-charge region) is actually linked to the gate potential than expected from the simple theory, the remainder being linked to the source and drain potentials. In Fig. 2, the geometry of this situation is illustrated. For devices that are not too short, a straightforward geometrical argument can be made from Fig. 2 in order to estimate the actual threshold voltage [54,56,62–73]. The depletion charge induced by the gate voltage is taken to lie approximately within the trapezoidal area of height W and lengths L and L_1, where L_1 is at the substrate side of the gate depletion region. The size of this charge is then approximately (charge per unit area)

$$\rho_{\text{depl}} = eWN_a(L + L_1)/2L. \tag{25}$$

This is then the depletion charge that must be induced by the gate at threshold, and if L is large, L_1 approaches L to give the long-channel results. For short channels, L_1 can be significantly smaller than L and, in fact, $L_1 \to 0$ as punchthrough occurs, so that punchthrough currents can actually occur well below the surface [72]. The drain depletion width in the corner region is given by

$$r_2 - r_j = (2\epsilon V_{\text{DB}}/eN_a)^{1/2}, \tag{26}$$

where $V_{\text{DB}} = V_D - V_B + V_{\text{bi}}$, V_{bi} is the built-in potential of the drain junction, r_2 the drain depletion region boundary radius, and r_j the drain metallurgical junction radius. A simple geometrical argument then gives

$$\rho_{\text{depl}}/\rho_L = 1 - (r_j/L)(g - 1) = f, \tag{27}$$

where ρ_L is the long-channel value and

$$g = \left\{ 1 + \frac{2}{r_j} \left(\frac{2\epsilon V_{\text{DB}}}{eN_a} \right)^{1/2} + \frac{1}{r_j^2} \left[\frac{2\epsilon V_{\text{DB}}}{eN_a} - \frac{2\epsilon(V_s - V_B)}{eN_a} \right] \right\}^{1/2}, \tag{28}$$

where we have used Eqs. (26) and (6). Expression (27) usually appears under the simplifying assumptions of $V_D = 0$ and $r_2 = r_j + W$. The parameter f is a unique function of the device geometry and material characteristics. The threshold is then given by using Eq. (27) in Eq. (7), and

$$V_T = - \frac{2\phi_b k_B T}{e} + V_{\text{FB}} - \frac{f}{C_0} \left[2\epsilon N_a e \left(V_C - V_B - \frac{2\phi_b k_B T}{e} \right) \right]^{1/2}, \tag{29}$$

where $V_C = 0$ is normally taken $(V_D = 0)$.

The presence of a potential at the source and/or drain can lead to a reduction of the gate depletion charge and thus to a reduction of the threshold voltage. This effect also leads to a reduction of the threshold in short channels just due to the presence of the depletion regions. This effect is shown in Fig. 3. It is evident that threshold reduction in short-channel de-

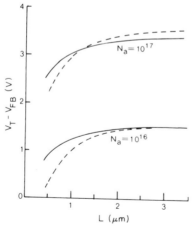

Fig. 3. Reduction in threshold voltage in short-channel devices according to Eq. (29). A substrate bias of -3 V is assumed. The solid curves are for $V_D = 0$ and the dashed curves are for $V_D = 3$ V.

vices becomes a significant effect that must be considered. Further, this reduction in threshold is dependent on the drain–source potential. If $V_{DB} \gg V_s$, then

$$g \approx 1 + \frac{(2\epsilon V_{DB}/eN_a)^{1/2}}{r_j} \qquad (30)$$

and

$$f \approx 1 - \frac{(2\epsilon V_{DB}/eN_a)^{1/2}}{L}. \qquad (31)$$

In this case, the dominant threshold shift is drain induced or, alternatively, substrate-bias induced. Within the approximation, this corresponds to the case $r_2 = r_j + W$ as well. In general, it has been suggested that the threshold voltage in short channels can be found from the simple relation [68]

$$V_T = V_{T,LC} - \alpha - \beta V_D, \qquad (32)$$

where α and β can be evaluated, for example, from two-dimensional calculations.

The threshold reduction in short-channel devices and its dependence on the various potentials such as drain and source make accurate control of the device threshold imperative. Thus turn-on stability over a long period of operation is crucial [74–76], and accurate knowledge of interface states and substrate homogeneity is necessary in fabrication [77,78]. Temperature stability also becomes important.

B. Subthreshold Conduction

One of the basic limitations of size reduction in VLSI is the spacing of the source and drain diffusions. Drain potential reverse biases the drain diffusion p–n junction and creates a field pattern that can lower the barrier separating it from the source. As the barrier is lowered, additional subthreshold current can flow, and the source–substrate–drain structure acts qualitatively similar to a bipolar transistor. There are two contributions to this drain-induced barrier lowering, as can be seen in Fig. 4. When the drain and source are at the same potential ($V_D = 0$), the potential under the gate may be symmetric, but, as discussed in the previous section, the separation between source and drain is such that the two depletion regions overlap; a weak punchthrough condition exists. The proximity effect of the two depletion regions leads to a lowering of the barrier between them, and enhanced diffusion current can flow. If a drain poten-

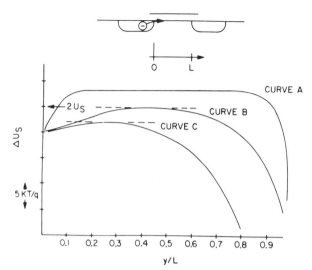

Fig. 4. Surface potential distribution for constant applied gate voltage (here we take $V_G = 1.8$ V). Only the channel length and the drain voltage are varied: For curve A, $L = 6.25$ μm and $V_{DS} = 0.5$ V; for curve B, $L = 1.25$ μm and $V_{DS} = 0.5$ V; for curve C, $L = 1.25$ μm and $V_{DS} = 5$ V. (After Troutman [72].)

tial is then applied, the drain depletion region spreads, thereby further lowering the barrier [56,62–73]. The ultimate proximity of the two diffusions that can be used in small devices is a function of the tolerable subthreshold current as much as it is of the doping level, oxide thickness, and voltage levels. If we use Eq. (32) in Eq. (22), we can write the subthreshold current in the semiempirical form

$$I_{ST} = I_0 \exp[(\beta e V_G/k_B T) + (\alpha e V_D/k_B T)], \qquad (33)$$

where the parameters α and β can be inferred from potential plots such as those of Fig. 4. In one such example, for a 2.1-μm device, Troutman and Chakravarti [65] found that α varied over the range 0.017–0.03. It can be expected that this value would increase considerably as the channel length is reduced.

Equation (33) just treats the surface component of the subthreshold current, however. Reference to Fig. 2 shows that as punchthrough is approached, the closest approach of the depletion regions lies along the path described by L_1 at the substrate end of the surface depletion region. It can be expected therefore that there will be a bulk, subsurface component of subthreshold current found [72]. In Figs. 5 and 6 we show the results of

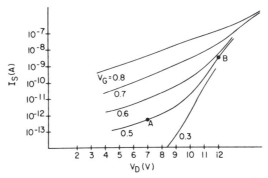

Fig. 5. Simulated IGFET low-level current–voltage characteristics for subthreshold currents with $T_{ox} = 40$ nm; $C_B = 4 \times 10^{15}$ cm^{-3}; $L = 1.8$ μm; $x_J = 2$ μm; $Y_J = 0.2$ μm; $V_x = 0$. (After Troutman [72].)

Troutman [72] that indicate that this is indeed the case. The geometry and specifications of the device were set so that punchthrough occurs prior to avalanche breakdown at the drain. Source–drain diffusions were modeled by a two-dimensional Gaussian profile of depth 2.0 μm and a lateral diffusion of 0.2 μm. A channel length of 1.8 μm and substrate doping of 4×10^{15} cm^{-3} were assumed for the two-dimensional computer simulations. The calculated subthreshold current–voltage plots are shown in Fig. 5, while in Fig. 6 are shown the current distributions in the channel. For the subthreshold current at point A (Fig. 5), electrons are injected over the surface potential barrier from the source, flow along the surface until they approach the drain, and then spread through the drain depletion region. They enter the drain below the surface, as is characteristic for current flow above pinch-off as well. Point B (Fig. 5) represents a point where the gate voltage dependence is weak and the drain voltage dependence is strong. Now most of the electrons are injected over the source potential barrier into the subsurface channel. The current density remains almost constant over the entire length of the channel and remains in a narrow subsurface conduction path. The difference between surface and subsurface conduction was first illustrated by Hachtel and Mack [26], but such a clear distinction between the two modes does not, in general, exist in FET structures. Rather, the two modes tend to merge together.

The subthreshold characteristics are important for utilizing devices in logic circuitry. However, simple linear scaling of a device is insufficient to hold subthreshold currents to an acceptable level. Many factors such as temperature, band gap, interface charge, and/or diffusion profiles are not scalable parameters but they have a profound influence on device performance, particularly in the subthreshold region.

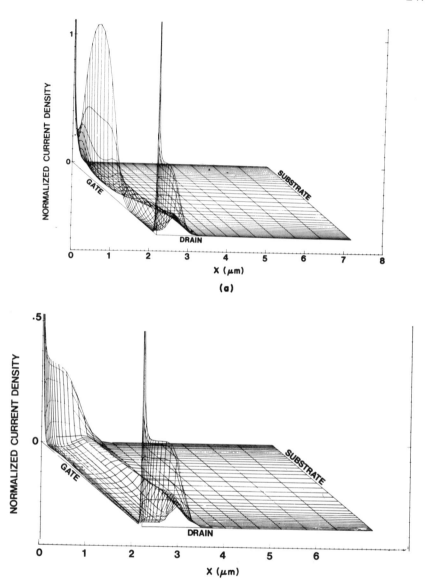

Fig. 6. Normalized current distribution in an IGFET: (a) curves for point A of Fig. 5; (b) curves for point B of Fig. 5. (After Troutman [72].)

C. Two-Dimensional Effects above Threshold

In the manner in which the MOSFET has developed, several short-channel effects have been observed that are not explicable from the normal gradual-channel approximation. We have previously discussed the variation of threshold voltage with channel length and the consequent exponential variation of subthreshold current with drain voltage. Both of these effects are a result of the two-dimensional sharing of the substrate depletion charge between the gate, source, and drain regions. There are, in addition, many consequences of this charge sharing above threshold as well, both in the pre- and post-saturation conduction characteristics of the device. The details of conduction in the saturation region were treated in the previous section. Short-channel effects arise from: (1) channel-length modulation, (2) charge sharing among the gate, source, and drain, (3) two-dimensional interaction between the junctions and the channel, and (4) reduced breakdown voltages. The first of these is the simple, effective, channel shortening introduced in the previous section. The second and third are related and, in fact, arise from the two-dimensional nature of the device. They can be treated in an approximate form by including the threshold voltage dependence on V_D in the above-threshold characteristics [70]. The last factor arises from lateral bipolar effects as will be discussed later.

The drain current arising from electron flow through the inversion channel is derived from Eq. (12) as

$$I_D = G_0[(V_G - V_T' - \tfrac{1}{2}V_D)V_D + (\bar{Q}_B/C_0)V_D]. \tag{34}$$

Equation (34) can be rewritten as

$$I_D = G_0(V_G - V_T'' - \tfrac{1}{2}V_D)V_D. \tag{35}$$

Equation (35) must still be modified by the effective threshold shift given by the factor f from Eq. (27) and the channel-length shortening. Thus the variation of the drain conductance with drain voltage in a short-channel device can be modeled by the variation of the threshold voltage with drain bias. The drain current is then

$$
\begin{aligned}
I_D &= \frac{Z\mu_e C_0}{L'} \left[\left(V_G - V_T'' - \frac{V_D}{2} \right) V_D \right] \\
&= \frac{Z\mu_e C_0}{L} \left[\left(V_G - V_T' - \frac{f\bar{Q}_B}{C_0} - \frac{V_D}{2} \right) V_D \right]
\end{aligned} \tag{36}
$$

for currents below saturation.

In the short-channel device, saturation is still assumed to occur when $\rho_{s,n}(L) = 0$. Extending Eq. (14) to the preceding case,

$$C_0(V_G - V_T' - V_{Dsat}) = f[2\epsilon e N_a(V_{Dsat} + V_{TB})]^{1/2}, \qquad (37)$$

where

$$f = 1 - \frac{r_j}{L} \left\{ \left[1 + \frac{2}{r_j} \left(\frac{2\epsilon V_{DB}}{eN_a} \right)^{1/2} \right. \right.$$
$$\left. \left. + \frac{1}{r_j^2} \left(\frac{2\epsilon V_{DB}}{eN_a} - \frac{2\epsilon(V_s - V_B)}{eN_a} \right) \right]^{1/2} - 1 \right\}, \qquad (38)$$

and $V_{DB} = V_D - V_B + V_{bi} = V_{Dsat} - V_B + V_{bi}$ for the present problem.

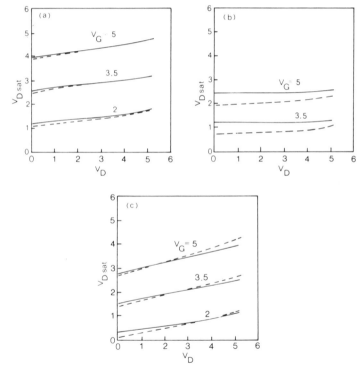

Fig. 7. Saturation voltage versus drain voltage V_D for a MOSFET with $y_0 = 250$ Å, $r_j = 0.3$ μm, and $V_{FB} = 0.6$ V for $V_B = 0$ (solid curves) and $V_B = -3$ (dashed curves): (a) $N_a = 10^{16}$ cm^{-3} and $L = 1.0$ μm; (b) $N_a = 5 \times 10^{16}$ cm^{-3} and $L = 1.0$ μm, where the curve is relatively flat as the parameters are such that V_T is relatively flat and insensitive to V_D; (c) $N_a = 5 \times 10^{16}$ cm^{-3} and $L = 0.5$ μm, where short-channel effects lead to a strong dependence of V_{Dsat} on V_D. A more negative V_{FB} would enhance these effects.

If we assume that f is a slowly varying function of V_D and take the approximate value at V_{Dsat}, then from Eq. (15),

$$V_{Dsat} = V_G - V_T + f^2 K^2 \{1 - [1 + (2V_G/K^2 f^2)]^{1/2}\}. \tag{39}$$

Thus we find that the actual saturation value of V_D is a function of V_D itself for short-channel devices. In Fig. 7, we show the variation of V_{Dsat} on V_D for a typical short-channel device. One view of Eq. (39) is that the drain saturation voltage rise and consequent drain current in saturation

$$I_{Dsat} = \frac{Z\mu_e C_0 V_{Dsat}{}^2}{L - [2\epsilon(V_D - V_{Dsat})/eN_a]^{1/2}} \tag{40}$$

are strongly affected by the drain voltage itself, due to a shift of the threshold voltage induced by the drain voltage, the latter shift being a consequence of drain-induced barrier lowering. The factor f is the primary variational factor here, although it is partially offset by the increase in the effective bulk charge $\bar{Q}_B$ and hence in K and can have a large variation from 0.1 to 1.0 in submicron devices. Clearly, the effect of the electrons being swept out of the channel into the bulk near the drain end, resulting in channel shortening, is also a factor.

IV. CARRIER INJECTION AND BREAKDOWN

With the high electric fields present in small devices, it is possible to have hot-carrier injection from the device into the oxide, leading to oxide charging and threshold voltage shifts [79–94]. The hot electron emission process in n-channel MOSFETs has been studied in some considerable detail and generally arises from two regions: either from the surface channel current or from the substrate leakage current.

The sources of hot electrons from the channel are the conduction electrons in the inverted channel and the multiplication current initiated by the channel current. Electrons flowing along the channel gain energy from the high fields of the drain depletion region after passing the pinch-off point. Those electrons that reach the interface and have sufficient energy to surmount the potential barrier of the oxide are emitted into the oxide. Even in the postsaturation region, for which $V_D > V_G$ and for which the field in the oxide near the drain would tend to return the electrons to the semiconductor, the electrons can remain in the oxide long enough to be trapped. Even if V_G is less than V_T, weak multiplication currents generated by the subthreshold current, or even by thermally generated carriers, may be sufficient to contribute a significant number of injected hot electrons to affect device performance.

The sources of substrate hot electrons are thermally generated leakage currents and weak carrier generations that arise from those currents in the drain depletion region. Electrons generated by these mechanisms, which are near the neutral depletion regions at the surface, can be accelerated toward the interface by the surface fields. Those that gain an energy greater than the oxide potential barrier are subsequently injected into the oxide [81,91].

These two sources for hot-carrier injection have been examined rather carefully by Ning *et al.* [91], for example. Their results suggest that this mechanism can be an important limitation for size reduction of MOSFETs. Results obtained on 1.0-μm devices, with $y_0 = 350$ Å and a substrate of 0.5-Ω cm material, suggest that the maximum operating voltage for such devices should be held to less than 3.5 V at 77 K and 4.75 V at 300 K [91]. In the dynamic mode of operation, voltages well above the source voltages can be achieved so that the supply source voltages must be held well below these limits.

The drain breakdown voltage due to avalanching in the high-field region of the depletion layer is also affected by short-channel effects. In particular, the drain breakdown voltage is observed to decrease considerably as the gate is biased into the turn-on condition. These results can be reasonably accounted for by taking a parasitic, lateral, bipolar action into account [58]. The hole current generated by weak avalanching in the drain depletion region flows back into the substrate region between source and drain. This current develops a voltage across the resistive substrate that tends to forward bias the source–substrate n–p junction, resulting in further charge injection into the substrate (base). Thus the source–substrate–drain acts like an n–p–n bipolar transistor and the drain breakdown voltage is reduced by a factor corresponding to $(h_{FE})^{1/2}$ of such a bipolar device. The current gain h_{FE} is proportional to $1/L$ so that the gain, and thus the breakdown voltage reduction, is affected as the channel length is reduced.

V. CARRIER TRANSPORT

In order to determine the current response accurately, one must solve an appropriate transport equation, as it is in these transport equations that many of the major modifications arise in the MOSFET. In the inversion layer, especially in strong inversion, the carriers are pulled up to the interface. These carriers are thus subject to modifications arising from channel quantization and additional scattering mechanisms. Also, the short time and space scales inherent in submicron devices can lead to modifications of the transport equations themselves. In this section, we want to examine these various aspects of inversion layer transport. First, we shall examine the nature of the Si–SiO$_2$ interface and then turn to the channel mobility. A detailed discussion of the quasi-two-dimensional transport is deferred to Chapter 7 of this volume, however. Finally, we shall discuss the retarded nature of the transport.

A. Nature of the Interface

The interface between Si and SiO$_2$ has been extensively investigated in an attempt to determine its width, composition, source of surface states, and effect on inversion layer transport. It is physically possible for crystalline silicon to terminate an amorphous SiO$_2$ film with a transition region that is only one bond-length wide [95]. However, there is evidence that the Si–SiO$_2$ interface is not a rigidly abrupt junction but that there is associated with it a bridging transition region in which the bonding changes from pure SiO$_2$ bonding to pure Si—Si bonding. The exact nature and width of this region is a topic of active debate and its nature has not been determined conclusively.

The general concensus now is that the interface is rough to 20 to 30 Å [96–99], although this may be due to undulations with an actual transition region of only 5 Å or less [100–104]. The actual interfacial region may contain an amorphous or disordered layer [105–107] or nonstoichiometric material [108–110]. However, the net result of this is that the interface is very sharp on the 10–30-Å scale atomically, a result consistent with the effect of the interface on scattering of carriers in the inversion layer [111]. This latter is important as the roughness of the surface interface leads to an additional scattering mechanism that lowers the mobility of the carriers in the inversion layer.

B. Channel Mobility

In the preceding discussion of the gradual-channel approximation, it was observed that the drain current, and hence the transconductance, was a function of the effective mobility of the electrons in the channel. The analysis of MOSFET properties has thus been sensitive to this parameter [112–114]. In general, it was discovered that the effective mobility is reduced in these devices when compared to the mobility of the bulk material. The results of studies of the inversion layer mobility suggests that additional surface-related scattering centers are affecting the mobility. These can arise from impurity scattering, surface roughness, crystal defects, interface states, and increased field transverse to the channel. Generally, the surface-effective mobility is found to decrease with increased inversion charge density, a result consistent with increased surface scattering in that the carriers are pushed closer to the surface/interface itself. In Fig. 8 we show the effective channel mobility as a function of the surface carrier density. One could also study the mobility as a function of the effective surface electric field, in which the latter is found by averaging the surface-normal field over the inversion layer, and [115]

$$E_{\text{eff}} = e(N_a W + \tfrac{1}{2} N_s)/\epsilon, \tag{41}$$

where N_s is the inversion charge density per unit area. By taking such an average, it is found that such diverse effects as tayloring the channel by ion implantation and applying substrate bias seem to fit the same mobility dependence [115]. The general falloff of mobility with carrier density as indicated in Fig. 8, would preclude interface charge scattering [116], as an opposite effect would be expected there. Rather, the mobility is most likely dominated by acoustic modes, both bulk and surface [117–120],

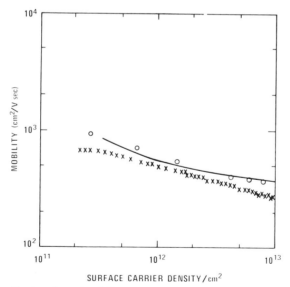

Fig. 8. The effective channel mobility as a function of the inversion carrier density at the (100)-Si face with a temperature of 300 K.

and by surface roughness scattering [121–123]. These shall be explored in much greater detail in Chapter 7 of this volume.

The relevance of hot carriers to the characteristics of MOSFETs became clear when the typical size of the devices began to become small. For example, a device of 10-μm channel length operating at 5 V has an average field of 5 kV/cm in the channel, and this field can heat the electrons far above thermal equilibrium. The first evidence for the importance of hot electrons for MOSFET characteristics was due to Hofstein and Warfield [48], but hot electron effects had earlier been suggested as the mechanism of current saturation [124]. Several studies have subsequently incorporated a field-dependent mobility in device analysis [125–132], even to incorporating dynamic carrier relaxation [133,134], and an extensive experimental study of high-field effects in MOSFETs has been performed [135]. In comparison with bulk silicon, hot electron effects seem smaller and occur at higher electric fields, although the resultant saturated velocity is not considerably less. This is shown in Fig. 9. It is currently felt that the saturated velocity is dominated by bulk intervalley phonons and that the weaker hot electron effects and lower value of the saturated velocity are a result of the lower surface mobility [136–139]. This will be discussed extensively in Chapter 7 of this volume.

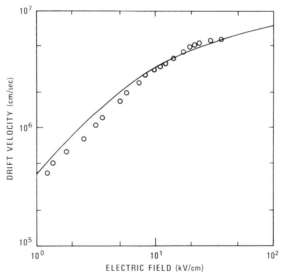

Fig. 9. The typical drift velocity for electrons in a (100)-Si surface inversion layer for $n_s = 6 \times 10^{12}$/cm^2 at 300 K. The solid curve is a theoretical fit and the data points are from Fang and Fowler [135].

C. Retarded Transport

Many of the problems that arise are due to the very fast time scales and small spatial scales inherent in small devices. For example, an electron traveling at 10^7 cm/sec can cross a 0.1-μm length in 10^{-12} sec. On this time scale, the electrons encountering a high-field region of this dimension, such as the pinch-off region in a MOSFET, do not have adequate time to establish any sort of equilibrium distribution [140–142]. Additional complications arise from the fact that the collision duration is no longer negligible on this time scale and can strongly affect the transport dynamics [142–146] and from the small spatial extent of the inversion layer leading to quantum effects in the energy spectrum of the carriers [60,61,147].

1. Balance Equations

The overriding theoretical concern in high-field transport in semiconductors is primarily one of discerning the form that the distribution function takes in the presence of the electric field [148]. However, for transport purposes, this is not an end product since integrals must be carried out over the distribution function in order to evaluate the transport coefficients [149]. The Boltzmann transport equation (BTE) has traditionally been used to find the form of the distribution function. Due to the very complicated nature of the BTE (when detailed scattering mechanisms are included it becomes essentially a nonlinear integrodifferential equation), it is usually not possible to solve it analytically and several possible assumptions can be made in order to obtain both $f(E)$ and its integrals [148–152]. However, in small semiconductor devices, the time scales are such that the use of the BTE must be questioned [143,144]. The problem arises from the fact that time scales on the order of the mean free times and the nonzero collision duration must be considered in these devices.

In traditional and semiclassical approaches to solutions of the BTE, it is assumed that the response of the carriers to the applied force is simultaneous with the applied force, even though the system may undergo subsequent relaxation. However, on the short time scale extant in submicron devices, a truly causal theory introduces memory effects that lead to convolution integrals in the transport coefficients [153–157]. Further, the collision terms in the BTE assume point collisions that occur instantaneously, which is fine when the mean time between collisions is large or when the collision duration is on a negligible time scale. At high fields, this is no longer the case, and correction terms must be generated to account for the retarded nature of the transport and for the nonzero time duration of the collision.

By utilizing a full quantum transport theory developed from the quantum-mechanical Liouville equation, a quantum kinetic equation (QKE), which incorporates the leading correction terms to the BTE, has been developed [158,159]. From this, moment equations have been developed in a straightforward manner [145,160]. Although a full path-variable solution of the QKE has been obtained [146], the moment equations are more straightforward in that their structure is formally analagous to the Langevin equation and the various terms have simple phenomenological meanings. These equations have the general form of [160]

$$\frac{3}{2} \frac{\partial}{\partial t} (k_B T_e) = e\mathbf{F} \cdot \mathbf{v}_d(t) - \frac{1}{\bar{\tau}_c} \int_0^t \exp\left(-\frac{\tau}{\bar{\tau}_c}\right) \langle \Gamma_E(t - \tau) \rangle \, d\tau \qquad (42)$$

for the energy balance equation and

$$m^* \frac{\partial v_d}{\partial t} = e\mathbf{F} - \frac{m^*}{\bar{\tau}_c} \int_0^t \mathbf{v}_d(t - \tau) \int_0^t \exp\left(-\frac{\tau'}{\bar{\tau}_c}\right) \langle \dot{\Gamma}_m(\tau - \tau') \rangle \, d\tau' \, d\tau \qquad (43)$$

for the momentum balance equation, where $\bar{\tau}_c$ is a mean collision duration, Γ_E the time rate of energy loss, and Γ_m the reciprocal momentum relaxation time.

There is an additional complication inherent in these equations that is not readily apparent from their structure. It was stated that the BTE is valid only in the weak coupling limit under the assumptions that the electric field is weak and slowly varying at most, the field and scatterers are independent, and the collisions occur instantaneously in space and time. Each of these approximations can be expected to be violated in future submicron-dimensioned semiconductor devices. At high fields, such as will occur in very small devices, the collision duration $\bar{\tau}_c$ is significant and correction terms must be generated (for the BTE) to account for this. These terms are, in fact, included in the QKE, and an additional field contribution appears as a differential superoperator term [158,159] (in the Liouville equation that generates the QKE) and, for collision integrals evaluated in the momentum representation, results in an intracollisional field effect [161–163]. This effect arises from the physical requirement of the electron being accelerated by the electric field during the collision. This effect modifies the matrix elements that appear in Γ_m and Γ_E in Eqs. (42) and (43) and is a direct consequence of the nonzero collision duration. In Fig. 10, the steady-state v–E curve is shown for bulk Si showing the intracollisional field effect. In Fig. 11, the transient velocity of electrons in silicon is shown for a homogeneous field of 20 kV/cm, applied at $t = 0$. The importance of the various effects contributing to collisional retardation is evident from this figure. The transient lasts for perhaps 0.5 psec at this field, and the effects are significant over a range of perhaps 0.1 μm, as

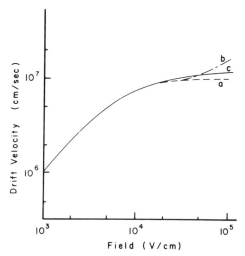

Fig. 10. The steady-state velocity vesus electric field curve in silicon at 300 K. Curve a is the normal semiclassical curve for Si, while curve b includes the intracollisional field effect. In curve c, full retarded solutions are used.

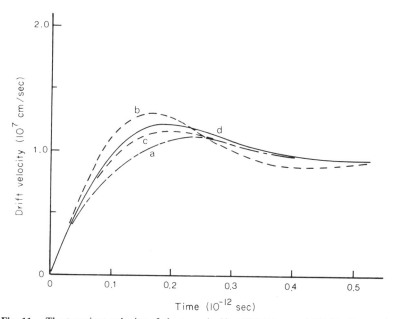

Fig. 11. The transient velocity of electrons in Si at 20 kV/cm and 300 K. Curve a is the normal transient response in the absence of retardation. Curve b includes the retardation due just to the finite collision duration, while curve c includes just the nonlocal response corrections. Curve d includes both effects.

suggested at the start of this subsection. The role of these effects within the operation of a submicron MOSFET has not yet been determined.

2. Diffusion

When the electrons reach the pinch-off point at the end of the channel in a MOSFET beyond saturation, they are injected into the high-field region of the drain depletion region. In this region, they move by a combination of drift and diffusion, but the latter tends to dominate the transport [164]. However, the current density assumes the presence of a valid Einstein relation when written as [164]

$$J = -eD_e n_s \, \nabla \phi_n = -k_B T_e \mu_e n_s \, \nabla \phi_n. \tag{44}$$

As previously discussed, transient relaxation events occur over a physical region that may approach $0.1 \ \mu$m, which is comparable to this high-field region. Inherent in the retarded transport in this region is another problem.

The derivation and validity of an Einstein relation rests on the presence of a stationary, steady-state, ensemble distribution function [165,166]. Stationarity here is in the stochastic sense that the velocity correlation between $v(t_1)$ and $v(t_2)$ is a function of $|t_2 - t_1|$. As evident from Fig. 11, a steady state certainly is not present in this region and the stationarity has not been established either, especially as the time scale is comparable to the nonzero collision duration. In Fig. 12 we show the diffusion in Si as

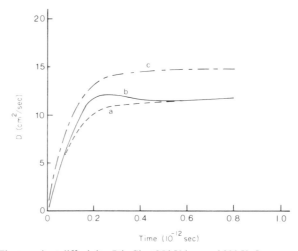

Fig. 12. The transient diffusivity D in Si at $25 \ \text{kV/cm}$ and $300 \ \text{K}$. In curve a, the standard $D = \langle (x - x_0)^2 \rangle / 2t$ was used, while in b, a retarded form was assumed [166]. Curve c was calculated from the transient mobility assuming that an Einstein relation, normally defined, could be used. All curves were calculated by a Monte Carlo technique.

calculated from $\langle (x - x_0)^2 \rangle / 2t$ and the retarded equivalent $\langle (x - x_0)^2 \rangle / 2t[1 - (1 - e^{-\gamma t})/\gamma t]^{-1}$ [166] and as calculated assuming that an Einstein relation holds, all calculated from a Monte Carlo technique. The lack of agreement among these illustrates the problem of assuming an Einstein relation exists. Further, the definitions used for the direct calculation of D assumes stationarity, a property that has not yet been established on the time scales of submicron devices. Why worry about D? In addition to the questions of device modeling, D is also important for estimates of the noise performance of many devices of interest. It is clear here that care must be used in using the basic phenomenological concepts of mobility and diffusion, especially in the submicron device regime.

REFERENCES

1. F. Braun, *Ann. Phys. Pogg.* **153**, 556 (1874).
2. J. C. Bose, U.S. Patent 775,840 (1904).
3. W. Schottky, *Naturwissenschaften* **26**, 843 (1938).
4. J. E. Lilienfeld, U.S. Patent 1,745,175 (1930).
5. O. Heil, British Patent 439,457 (1935).
6. W. Shockley and G. L. Pearson, *Phys. Rev.* **74**, 232 (1948).
7. J. L. Moll, *IEE Wescon Conv. Rec. Part 3* 32 (1948).
8. W. G. Pfann and G. C. B. Garrett, *Proc. IRE* **47**, 2011 (1959).
9. D. Kahng and M. M. Atalla, *Proc. Solid-State Device Res. Conf.* unpublished, (1960).
10. D. R. Frankl, *Solid-State Electron.* **2**, 71 (1961).
11. R. Lindner, *Bell Syst. Tech. J.* **41**, 803 (1962).
12. L. M. Terman, *Solid-State Electron.* **5**, 285 (1962).
13. K. Lehovec, A. Slobodsky, and J. L. Sprague, *Phys. Status Solidi* **3**, 447 (1963).
14. H. K. J. Ihantolla and J. L. Moll, *Solid-State Electron.* **7**, 423 (1964).
15. J. E. Johnson, *Solid-State Electron.* **7**, 861 (1964).
16. C. T. Sah, *IEEE Trans. Electron Devices* **ED-11**, 324 (1964).
17. S. R. Hofstein and F. P. Heiman, *Proc. IEEE* **51**, 1190 (1963).
18. H. G. Dill, *Sci. Electr.* **13**, 45 (1967).
19. T. Hayashi, Y. Tarui, and T. Komuro, *Bull. Electrotech. Lab.* **31**, 7 (1967).
20. J. Grosvalet, *Onde Electr.* **47**, 937 (1967).
21. W. W. Lattin, *Proc. Asilomar Conf. Circuits, Syst., Comput., 8th* (S. R. Parker, ed.), p. 23. IEEE Press, New York, 1975.
22. F. M. Klaasen, *Proc. NATO ASI Process Device Modeling Integrated Circuit Design* (F. van de Wiele, W. L. Engl, and P. G. Jespers, eds.), p. 541. Noordhoff, Groningen, The Netherlands, 1977.
23. M. Reiser and P. Wolf, *Electron. Lett.* **8**, 254 (1972).
24. F. H. De la Moneda, *IEE Trans. Circuit Theory* **CT-20**, 666 (1973).
25. M. Reiser, *IEEE Trans. Electron Devices* **ED-20**, 35 (1973).
26. G. D. Hachtel and M. H. Mack, *Int. Solid State Circuits Conf. Digest* p. 110 (1973).
27. J. J. Barnes and R. J. Lomax, *Electron. Lett.* **10**, 341 (1974).
28. P. E. Cottrell and E. M. Buturla, *Proc. Int. Electron Devices Meeting* p. 51. IEEE Press, New York, 1975.

29. W. Fichtner, *IEEE J. Solid State Electron Devices* **2**, 47 (1978).
30. D. B. Scott and S. G. Chamberlin, *IEEE J. Solid State Circuits* **SC-14**, 633 (1979).
31. H. N. Kotecha and K. E. Beilstein, *Proc. Int. Electron Devices Meeting* p. 47. IEEE Press, New York, 1975.
32. K. E. Kroll and G. K. Ackermann, *Solid State Electron.* **19**, 77 (1976).
33. F. H. Gaensslen, *Proc. Int. Electron Devices Meeting* p. 512. IEEE Press, New York, 1977.
34. J. S. T. Huang, *IEEE Trans. Electron Devices* **ED-20**, 513 (1970).
35. J. R. Edwards and G. Marr, *IEEE Trans. Electron Devices* **ED-20**, 283 (1970).
36. J. Borel, J. Bernard, and J.-P. Suat, *Solid State Electron.* **12**, 1377 (1973).
37. J. A. Sauvage and S. A. Evans, *Proc. Int. Electron Devices Meeting* p. 61. IEEE Press, New York 1973.
38. V. L. Rideout, F. H. Gaensslen, and A. Leblanc, *Proc. Int. Electron Devices Meeting* p. 148. IEEE Press, New York, 1973.
39. E. C. Douglas and A. G. G. Dingwall, *IEEE Trans. Electron Devices* **ED-21**, 324 (1974).
40. J. S. T. Huang, *IEEE Trans. Electron Devices* **ED-21**, 995 (1974).
41. H. Runge, *Electron. Eng.* **48**, 41 (1976).
42. B. Farzan and C. A. T. Salama, *Solid State Electron.* **19**, 297 (1976).
43. T. J. Rodgers, S. Asai, M. D. Pocha, R. W. Dutton, and J. D. Meindl, *IEEE J. Solid State Circuits* **SC-10**, 322 (1975).
44. I. Yoshida, T. Musuhara, M. Kubo, and T. Tokuyama, *Oyo Buturi* **44**, 249 (1975).
45. M. D. Pocha and R. W. Dutton, *IEEE J. Solid State Circuits* **SC-11**, 718 (1976).
46. H. C. Card and E. L. Heasell, *Solid-State Electron.* **19**, 965 (1976).
47. D. Frohman-Bentchkowsky, *IEEE J. Solid State Circuits* **SC-6**, 301 (1976).
48. S. R. Hofstein and G. Warfield, *IEEE Trans. Electron Devices* **ED-12**, 129 (1965).
49. J. A. Guerst, *Solid State Electron.* **9**, 129 (1966).
50. J. E. Schroeder and R. S. Muller, *J. Appl. Phys.* **45**, 828 (1974).
51. V. G. K. Reddi and C. T. Sah, *IEEE Trans. Electron Devices* **ED-12**, 139 (1965).
52. G. Baum and H. Beneking, *Arch. Elektrotech. Uebertragung* **22**, 1 (1968).
53. D. Frohman-Bentchkowsky and A. S. Grove, *IEEE Trans. Electron Devices* **ED-16**, 108 (1969).
54. D. P. Smith, Thesis, Stanford Univ., unpublished (1971).
55. V. A. Fesechko and N. B. Grudanow, *Izv. Vuz Radioelektron.* **21**, 134 (1978).
56. R. H. Dennard, F. H. Gaensslen, H. N. Yu, V. L. Rideout, E. Bassous, and A. R. Leblanc, *IEEE J. Solid State Circuits* **SC-9**, 256 (1974).
57. F. H. Gaensslen, V. L. Rideout, and E. J. Walker, *Proc. Int. Electron Devices Meeting* p. 43. IEEE Press, New York, 1975.
58. H. Masuda, M. Nakai, and M. Kubo, *IEEE Trans. Electron Devices* **ED-26**, 980 (1979).
59. B. Hoeneisen and C. A. Mead, *Solid State Electron.* **15**, 819 (1972).
60. F. Stern, *Crit. Rev. Solid State Sci.* **4**, 499 (1974).
61. F. Stern and W. E. Howard, *Phys. Rev.* **163**, 816 (1967).
62. R. R. Troutman, *Digest 1973 Solid-State Circuits Conf.* p. 108 (1973).
63. H. S. Lee, *Solid State Electron.* **16**, 1407 (1973).
64. L. D. Yau, *Solid State Electron.* **17**, 1509 (1974).
65. R. R. Troutman and S. N. Chakravarti, *IEEE Trans. Circuit Theory* **CT-20**, 659 (1973).
66. R. R. Troutman, *IEEE J. Solid State Circuits* **SC-9**, 55 (1974).
67. W. R. Bandy and D. P. Kokalis, *Solid State Electron.* **20**, 675 (1977).

68. R. R. Troutman and A. G. Fortino, *IEEE Trans. Electron Devices* **ED-24,** 1266 (1977).
69. G. W. Taylor, *IEEE Trans. Electron Devices* **ED-25,** 337 (1978).
70. G. W. Taylor, *Solid State Electron.* **22,** 701 (1979).
71. G. R. M. Rao, *Solid State Electron.* **22,** 729 (1979).
72. R. R. Troutman, *IEEE Trans. Electron Devices* **ED-26,** 461 (1979).
73. W. Fichtner and H. W. Pötzl, *Int. J. Electron.* **46,** 33 (1979).
74. M. H. White and J. R. Cricchi, *Solid State Electron.* **9,** 991 (1966).
75. F. H. Reynolds, *Proc. IEEE* **119,** 1683 (1972).
76. F. Fischer and J. Fellinger, *Siemens Forsch. Entwick.* **5,** 25 (1976).
77. R. S. Cobbold, *Electron. Lett.* **2,** 109 (1966).
78. S. D. Brotherton, *Solid State Electron.* **10,** 611 (1967).
79. H. Nara, Y. Okamoto, and H. Ohnuma, *Jpn. J. Appl. Phys.* **9,** 1103 (1970).
80. S. A. Abbas and R. C. Dockerty, *Appl. Phys. Lett.* **27,** 147 (1975).
81. T. H. Ning, C. M. Osburn, and H. N. Yu, *Appl. Phys. Lett.* **29,** 198 (1976).
82. A. H. Phillips, Jr., R. R. O'Brien, and R. C. Joy, *Proc. Int. Electron Devices Meeting* p. 39. IEEE Press, New York, 1975.
83. S. A. Abbas and R. C. Dockerty, *Proc. Int. Electron Devices Meeting* p. 35. IEEE Press, New York, 1975.
84. T. H. Ning, C. M. Osburn, and H. N. Yu, *J. Electron. Mater.* **6,** 65 (1976).
85. D. R. Young, *J. Appl. Phys.* **47,** 2098 (1976).
86. R. R. Troutman, *Proc. Int. Electron Devices Meeting* p. 578. IEEE Press, New York, 1976.
87. T. H. Ning, C. M. Osburn, and H. N. Yu, *J. Appl. Phys.* **48,** 268 (1977).
88. T. H. Ning, *Proc. Int. Electron Devices Meeting* p. 144. IEEE Press, New York, 1977.
89. T. H. Ning, *Solid State Electron.* **21,** 273 (1978).
90. R. R. Troutman, *Solid State Electron.* **21,** 283 (1978).
91. T. H. Ning, P. W. Cook, R. H. Dennard, C. M. Osburn, S. E. Schuster, and H. N. Yu, *IEEE Trans. Electron Devices* **ED-26,** 346 (1979).
92. P. E. Cottrell, R. R. Troutman, and T. H. Ning, *IEEE Trans. Electron Devices* **ED-26,** 520 (1979).
93. P. K. Chatterjee, *Proc. Int. Electron Devices Meeting* p. 14. IEEE Press, New York, 1979.
94. C. Hu, *Proc. Int. Electron Devices Meeting* p. 22. IEEE Press, New York, 1979.
95. S. T. Pantelides, *J. Vac. Sci. Technol.* **14,** 965 (1977).
96. S. I. Raider, R. Flitsch, and M. J. Palmer, *J. Electrochem. Soc.* **122,** 413 (1922).
97. R. A. Clarke, R. L. Trapping, M. A. Hopper, and L. Young, *J. Electrochem. Soc.* **122,** 1347 (1975).
98. W. L. Harrington, R. E. Honig, A. M. Goodman, and R. Williams, *Appl. Phys. Lett.* **27,** 644 (1975).
99. C. R. Helms, W. E. Spicer, and N. M. Johnson, *Solid-State Commun.* **25,** 673 (1978).
100. J. S. Johannessen, W. E. Spicer, and Y. E. Strausser, *J. Appl. Phys.* **47,** 3028 (1976).
101. L. C. Feldman, P. J. Silverman, and I. Stensgaard, *Proc. Int. Conf. Phys. SiO₂ Interfaces, Yorktown Heights, New York* (S. Pantelides, ed.), p. 344. Pergamon, Oxford, 1978.
102. J. Blanc, C. J. Buiocchi, M. S. Abrahams, and W. E. Ham, *Appl. Phys. Lett.* **30,** 120 (1977).
103. O. L. Krivanek, T. T. Sheng, and D. C. Tsui, *Appl. Phys. Lett.* **32,** 437 (1978).
104. J. F. Wager and C. W. Wilmsen, *J. Appl. Phys.* **50,** 874 (1979).
105. C. C. Change, P. Petroff, G. Quintana, and J. Sosniak, *Surf. Sci.* **38,** 341 (1978).

106. T. W. Sigmon, W. K. Chu, E. Lugujjo, and J. W. Mayer, *Appl. Phys. Lett.* **24,** 105 (1974).
107. P. Offerman, *J. Appl. Phys.* **48,** 1890 (1977).
108. S. I. Raider and R. Flitsch, *Proc. Int. Conf. Phys. SiO₂ Interfaces, Yorktown Heights, New York* (S. Pantelides, ed.), p. 344. Pergamon, Oxford, 1978.
109. F. J. Grunthaner, P. J. Grunthaner, R. P. Vasquez, B. F. Lewis, and J. Maserjian, *J. Vac. Sci. Technol.* **16,** 1443 (1979).
110. D. E. Aspnes and J. B. Theeten, *Phys. Rev. Lett.* **43,** 1046 (1979).
111. A. Hartstein, T. H. Ning, and A. B. Fowler, *Surf. Sci.* **58,** 178 (1976).
112. O. Leistiko, Jr., A. S. Grove, and C. T. Sah, *IEEE Trans. Electron Devices* **ED-12,** 248 (1965).
113. R. Y. Deshpande, *Solid-State Electron.* **8,** 313 (1965).
114. R. F. Pierret and C. T. Sah, *Solid-State Electron* **11,** 279 (1968).
115. A. G. Sabnis and J. T. Clemens, *Proc. Int. Electron Devices Meeting* p. 18. IEEE Press, New York, 1979.
116. C. T. Sah, T. H. Ning, and L. L. Tschopp, *Surf. Sci.* **32,** 561 (1972).
117. S. Kawaji, *J. Phys. Soc. Jpn.* **27,** 608 (1967).
118. H. Ezawa, T. Kuroda, and K. Nakamura, *Surf. Sci.* **24,** 569 (1971).
119. H. Ezawa, S. Kawaji, and K. Nakamura, *Surf. Sci.* **27,** 218 (1971).
120. H. Ezawa, S. Kawaji, and K. Nakamura, *J. Appl. Phys.* **13,** 126 (1974).
121. J. R. Brews, *J. Appl. Phys.* **43,** 2306 (1976).
122. Y. C. Cheng, *J. Appl. Phys.* **44,** 2425 (1973).
123. D. K. Ferry, *Proc. Int. Conf. Phys. Semicond., 13th* p. 758. Tipo Graves, Rome, 1976.
124. J. Grosvalet, C. Motsch, and R. Tribes, *Solid-State Electron.* **6,** 65 (1963).
125. F. N. Trofimenkoff, *Proc. IEEE* **53,** 1765 (1965).
126. K. Tarnay, *Proc. IEEE* **54,** 1077 (1966).
127. G. F. Neumark, *Solid-State Electron.* **10,** 169 (1967).
128. J. Borel and G. Merckel, *Proc. Int. Conf. Adv. Microelectron,* Feder. Nat. Ind. Electr., Paris, 1970.
129. A. A. Molchanov and L. S. Khodosh, *Izv. Vuz Radioelectron.* **13,** 892, (1970).
130. B. Hoeneisen and C. A. Mead, *IEEE Trans. Electron Devices* **ED-19,** 382 (1970).
131. L. Szanto, *Tesla Electron.* **8,** 20 (1975).
132. R. W. Hockney, R. A. Warriner, and M. Rieser, *Electron. Lett.* **10,** 484 (1974).
133. B. Hoefflinger, H. Silbert, and G. Zimmer, *IEEE Trans. Electron Devices* **ED-26,** 513 (1979).
134. B. Carnez, A. Cappy, A. Kaszinski, and G. Salmer, *Proc. Eur. Microwave Conf., 8th* p. 410. Microwave Exhibition and Publ. Sevenoaks, England, 1978.
135. F. F. Fang and A. B. Fowler, *J. Appl. Phys.* **41,** 1825 (1970).
136. K. Hess and C. T. Sah, *J. Appl. Phys.* **45,** 1254 (1974).
137. K. Hess and C. T. Sah, *Phys. Rev. B* **10,** 3375 (1974).
138. D. K. Ferry, *Phys. Rev. B* **14,** 5364 (1976).
139. D. K. Ferry, *Solid-State Electron.* **21,** 115 (1978).
140. J. G. Ruch, *IEEE Trans. Electron Devices* **ED-19,** 652 (1972).
141. T. J. Maloney and J. Frey, *J. Appl. Phys.* **48,** 781 (1977).
142. D. K. Ferry and J. R. Barker, *Solid-State Electron.* **23,** 519 (1980).
143. J. R. Barker and D. K. Ferry, *Solid-State Electron.* **23,** 531 (1980).
144. D. K. Ferry and J. R. Barker, *Solid-State Electron.* **23,** 545 (1980).
145. D. K. Ferry and J. R. Barker, *Solid State Commun.* **30,** 361 (1979).
146. J. R. Barker and D. K. Ferry, *Phys. Rev. Lett.* **42,** 1779 (1979).

147. F. F. Fang and A. B. Fowler, *Phys. Rev.* **41**, 1825 (1970).
148. D. K. Ferry, *in* "Handbook of Semiconductors" (W. Paul, ed.), Vol. I. North-Holland, Amsterdam (in press).
149. M. Dresden, *Rev. Mod. Phys.* **33**, 265 (1961).
150. E. Conwell, "High Field Transport in Semiconductors." Academic Press, New York, 1967.
151. P. Price, *IBM J. Res. Dev.* **14**, 12 (1970).
152. P. Price, *Solid-State Electron.* **21**, 9 (1978).
153. R. Zwanzig, *J. Chem. Phys.* **40**, 2527 (1964).
154. M. S. Green, *J. Chem. Phys.* **20**, 1281 (1952); **22**, 398 (1954).
155. R. Kubo, *J. Phys. Soc. Jpn.* **12**, 570 (1957).
156. H. Mori, *Phys. Rev.* **112**, 1829 (1958).
157. R. Zwanzig, *Phys. Rev.* **124**, 983 (1961).
158. J. R. Barker, *J. Phys. C* **6**, 2663 (1973).
159. J. R. Barker, *in* "Handbook of Semiconductors" (W. Paul, ed.), Vol. 1. North-Holland Publ., Amsterdam (in press).
160. D. K. Ferry and J. R. Barker, *J. Phys. Chem. Sol.* **41**, 1083 (1980).
161. K. K. Thornber, *Solid-State Electron.* **21**, 259 (1978).
162. J. R. Barker, *Solid-State Electron.* **21**, 267 (1978).
163. D. K. Ferry, *in* "Nonlinear Transport in Semiconductors" (D. K. Ferry, J. R. Barker, and C. Jacoboni, eds.), pp. 117–125. Plenum, New York, 1980.
164. H. C. Pao and C. T. Sah, *Solid-State Electron.* **9**, 927 (1966).
165. R. Kubo, *Rep. Prog. Phys.* **29**, 255, (1966).
166. R. Kubo, *in* "Lecture Notes in Physics 31: Transport Phenomena" (J. Ehlers, K. Hepp, and H. A. Weidenmuller, eds.), p. 74. Springer-Verlag, Berlin and New York, 1974.

Chapter 7

The Government Role in VLSI

JAY H. HARRIS* *et al.*

Division of Electrical, Computer, and Systems Engineering
National Science Foundation
Washington, D.C.

I. INTRODUCTION

The government role in the development of VLSI is a complex one because VLSI itself can be perceived in several ways. From one perspective it is an active current research effort with clear-cut, long-term applications in electronic systems. From another perspective the term represents

* *Present address:* College of Engineering, San Diego State University, San Diego, California 92182

the widely acknowledged future direction of a major American industry. From yet another perspective it represents the cornerstone of the new computer/communications age that will see a pervasive use of intelligent electronics systems and may lead to major social and economic changes. It is probably safe to say that as a result of this spectrum of perceptions, every agency of the executive branch of the federal government and every congressional office and committee has some level of interest in VLSI.

Government organizations concerned with research and development in electronics and related areas have the most intense interest in VLSI. The Department of Defense (DoD) and related agencies see VLSI as important to their mission, and they have developed significant programs intended to hasten its development. This role is consistent with DoD's long term experience with electronics and its significant role in the development of integrated circuits. Other science and technology agencies, such as the National Bureau of Standards (NBS), the National Science Foundation (NSF), the National Aeronautics and Space Administration (NASA), The National Institutes of Health (NIH), and the Department of Energy (DoE), have their own quantitatively less intense interests in the field that reflect their own particular missions.

VLSI as a commercial interest can potentially impact a larger group of government organizations. The degree of concern of these organizations often reflects their level of interaction with trade associations and individual companies. For example, Congressman McCloskey of California has attributed recent changes in the nation's capital gains tax structure to activities of the semiconductor industry [1]. The activities were motivated in part by industrial concern about the availability of capital to pay the escalating cost of processing equipment. Those costs will doubtless continue to escalate in the VLSI era.

The semiconductor/computer industry is viewed by the government as a "lead industry" whose continuing success must not be impeded unnecessarily by governmental regulations and whose success is sufficiently important to the nation to justify government programs of support where they appear appropriate. A 1979 White House policy review [2] concerned with industrial innovation underscored this commitment to action, at least for generic technologies, i.e., for technologies that underlie broad industrial sectors. Although the policy review has been criticized for its limited scope and its emphasis on government programs, the review does enunciate the principle of government commitment to support of technological advance in healthy industries. A wide spectrum of future actions could evolve from the humble beginnings of the review. Jordan Baruch of the Department of Commerce, who spearheaded the study, has cited VLSI as a prime example of the kind of generic technology of concern in the review. Depending in part on industry initiative, new governmental efforts to help the electronics industry to maintain its commercial vitality could evolve, as could changes in tax, antitrust, and trade policies.

The government as a whole will impact and be impacted by whatever long term revolutionary social and economic effects are implied by VLSI. From labor issues to foreign affairs, from income tax collection to library management, and from incentive purchasing to occupational safety, there is a very extensive list of government functions and organizations that relate to future developments in VLSI. As with any national issue, the government structure that will deal with VLSI is directed by the Office of the President and by the Congress, which formally initiates and approves new programs and augments or reduces old programs. Within the White House, science-related policy is developed and coordinated by the Office of Science and Technology Policy (OSTP). Oversight agencies such as the Office of Management and Budget (OMB) in the executive branch and the General Accounting Office (GAO) in the congressional branch of government review the effectiveness of programs and interact with congressional committees and decision makers in the various agencies. There are a total of nineteen federal agencies that support research and development (R&D) programs. These agencies create and implement programs and coordinate their efforts at various levels. Interagency groups in electronics and in materials, for example, coordinate support of individual research grants and contracts related to VLSI.

This chapter cannot explore the full spectrum of the relationship between the government and VLSI. Using the format of contributed sections from agency participants and from policy analysts, a sampling is provided of past and future agency activities related to VLSI. The discussion begins with contributions from the agency that provides primary support for electronics.

II. DEPARTMENT OF DEFENSE PROGRAMS

A. VLSI and National Defense

George Gamota[1]

1. Background

Over the years, the Department of Defense has played a key role in developing or being the stimulant for whole new technologies having tremendous spin-offs in the commercial markets. Some examples include

---

[1] Office of the Undersecretary of Defense for Research and Engineering, Department of Defense, Washington, D.C.

synthetic rubber, nuclear power, the jet engine, the transistor, and integrated circuits.

Each of these technologies was motivated by a military requirement. In the case of the integrated circuits, for example, the Air Force needed to miniaturize its electronics. To promote a sense of smallness of electronic devices, the Air Force program was called "Molecular Electronics." In retrospect, it turns out that the integrated circuit (IC) that was developed was not of molecular level size nor was it under a DoD contract. However, it is clear that DoD's interest and available funding stimulated the thinking that led directly to the invention of the integrated circuit.

During the late 1950s, DoD provided nearly a quarter of the R&D funds to the semiconductor industry, not counting a large portion of the industrial Independent Research and Development (IR&D) effort. The main focus was on the miniaturization and reliability of electronic devices. Following the development of the first working integrated circuit by Jack Kilby at Texas Instruments (TI), under TI's own R&D effort, the Air Force funded TI in 1959 and 1960 at a level of $3.2 million for R&D in this area. The DoD's R&D effort continued, but clearly the potential military market ultimately was the driving force behind the large industrial interest in this area. For example, as early as 1962, or only four years after the invention of the IC, the devices were to be used on the Minuteman I guidance and control systems. At that time, DoD represented 100% of the market. Other agencies, most notably NASA, contributed to the IC development but that came later, and it never amounted to a significant fraction in those early years.

The value of integrated circuits defies enumeration, and as we continue to achieve higher and higher densities and shrink dimensions further, new and more exciting possibilities come into view. Today, DoD is but 7% of the IC market, since most markets are now directed to commercial applications. Today electronic devices using ICs can be found in every home, and in the next five to ten years, ICs wil penetrate virtually all phases of our lives.

Of course, DoD has greatly benefited from this thrust in the civilian market. The competition has driven the price to the point that the electronics—the computer on a chip—no longer is a significant factor in the cost of devices such as calculators, tv games, and electronic watches. By shrinking in size and price, the sales have grown; in 1977, the value of U.S. IC business was about $2.5 billion of which $300 million was in microprocessor sales. In 1980, microprocessor sales along will top the $1 billion mark.

The need for microminiaturization continues in DoD, and the goal is still to develop "molecular" electronics. Industry, however, in response to market forces, is, to a large degree, not pushing the frontier of size and speed. So again DoD is determined to use its direct funding of R&D to accomplish this goal. Three efforts will be discussed in this chapter: a near-term exploratory development effort, the very high speed integrated circuit (VHSIC) program described in Section II.B; a midrange VLSI program described in Section II.C; and a much longer term basic research endeavor, the ultrasmall electronics research program (USER) which will be described next.

2. Ultrasmall Electronics Research (USER)

The recently observed slowdown of industrial interest in further microminiaturization has been attributed to two factors, the lack of product definition and design and the lack of available manpower for the design/layout of circuits. While the latter reason is all-prevailing in the electronics job market, it can be argued that DoD has a clearly defined viewpoint of the definition and design of future electronic components which is dictated by the complexity of future military systems found in Electronic Warfare, Search and Surveillance, and Command, Communication, and Control (C^3) systems. These systems are unique to DoD, and the requirements can only be met by drastically reducing the processing time and power consumption while increasing reliability. Thus a specific, exclusive DoD initiative is required to provide the needed technology base for future electronics-oriented weapons systems.

Today progress in the microelectronics industry is strongly coupled with the ability to continue to make ever-increasing numbers of smaller and more sophisticated devices on a single chip; i.e., the move to very large scale integrated (VLSI) circuits. It is also apparent that extrapolation from present technology will produce individual devices whose dimensions are on the order of 0.2 to 0.5 μm. The new VHSIC program of the Department of Defense is addressing the regime from 0.7 to 1.2 μm. On the other hand, the advent of high-resolution electron-beam, x-ray, molecular, and ion-beam lithographic techniques is pushing toward an era of ultrasmall devices in which individual feature sizes might well be fabricated on the macromolecular scale; i.e., 10–20 nm. Some structures that have been proposed are so small that the bulk properties of the host semiconductor material may be significantly less important than size-related effects. Effects such as tunneling, size quantization, range order, and fluctuation phenomena, which are produced by the device surfaces and the interaction with neighboring structures, become important. In these devices, preliminary studies indicated that the temporal and spatial scales become sufficiently short and the electric fields sufficiently large that new physical regions are reached where the precepts of present-day, semiclassical device physics are inappropriate and, indeed, may be misleading. Moreover, the new physical properties that are available lead to the possibility of radically new electronic device structures in which the individual device may assume a variety of functions that depend on the influence of neighboring devices.

Electronics research in nanometer dimensions is required to develop the theoretical and experimental tools to handle the new physics involved. Research in ultrasmall dimensions deals mainly with physics, chemistry, metallurgy, and carrier transport in highly constrained geometrical structures, such as may be used for devices. The conventional electron device models no longer apply when dimensions shrink below 100 nm and new quantum-mechanical concepts for the electron transport have to be found. This new research area addresses material questions and questions that are raised by pushing for higher IC complexity and higher speeds than are presently available or will be available with current technology. Some researchers call ultrasmall electronic research one of the last remaining barriers of solid-state electronics, where the new fundamental unit is an aggregate or array of molecules or atoms. While this research is highly speculative, the potential payoff is very high, particularly for the next generation of VLSI and related applications, and even in nonelectronic areas such as chemical catalysis and microbiology.

The USER program is intended to be a coherent, long-range, multidisciplinary research program that addresses the scientific problems expected to arise in nanometer electronic structures. Many of the problems are inherently difficult because the small structural sizes are no longer larger than the fundamental electron path lengths involved in transport, as they are in more classical devices. The USER program will address important relevant aspects of the nucleation, growth, and properties of materials as well as electron transport properties in individual ultrasmall structures and in interacting ultrasmall structures. While the coupling of the structures to the macroscopic world is important and will be considered, much of the technological aspects, such as detailed lithographic or fabrication questions, will not be considered in the program as it is expected that these questions will be addressed within the NSF and by ongoing industrial programs. In the most simple terms the objective of the program is to generate a new science of ultrasmall domains.

Expanding on the areas to be addressed, we shall now briefly discuss the four types of research problems.

a. Nucleation. The nucleation of a thin film is governed generally by the detailed nonequilibrium thermodynamics of the system and includes questions of kinetics/kinematics of weak atomic bonding, molecular forces motion, and cluster formation. These detailed processes are not well understood. In the ultrasmall scale our current models for such effects as defect formation, chemisorption, segregation/agglomeration, atomic solubility, solid phase reactions, and microinclusions are inapplicable. Whereas certain effects may be second order or less in large systems, they may easily become the dominant effects on the ultrasmall scale.

b. Material Properties. The growth, crystallinity, and structure of thin material layers is pervasively important to USER. In an ultrasmall electronic structure, the device dimension is approaching the dimension of long-range order in the material. For this reason, we must be concerned with the adequacy of our current understanding of the concepts of diffusion/migration, microstructure, and phase transitions within the host material. Again, effects such as microinclusion or precipitate structures may be negligible in a large structure, but the physical dimension involved can well be expected to be a large fraction of the size of an ultrasmall device structure, and such effects will necessarily become dominant.

c. Transport Properties. The transport of carriers in the current *modus operandi* of semiconductor electronic devices cannot be expected to maintain that role in future ultrasmall electronic devices. Current approaches are based on a transport equation formalism, utilizing the Boltzmann transport equation. Future inquiries must seek alternative approaches, since almost all of the fundamental assumptions on which the Boltzmann equation is based can be expected to be violated in ultrasmall structures. The change in the electronic structure in dimensionally constrained systems alters all scattering rates and state-density functions, strongly modifying transport effects. Models of memory and quantum effects in transport, boundary scattering, nonequilibrium phonons, noncontinuum models (self-consistency requirements), screening, and dielectric response are inadequate on the small temporal and spatial scales envisioned in ultrasmall structures. New transport calculational methods are required for modeling of ultrasmall devices/structures.

d. Interacting Structures. On the ultrasmall scale, including the effects of structural boundaries and the presence of an array of structures in analytical approaches to the characterization of physical properties can lead to a renormalization of the lattice vibrational energy spectrum and the electron-carrier dynamics as well as to consideration of long-range dissipative interactions. Distributions reflecting the position, size, and time ordering of the array must be developed formally, on a mathematical basis, and the interaction mechanisms (electrical, magnetic, optical, acoustic, etc.) especially considered. The role of interstructure interactions leading to structural renormalization, synergetic self-organization, collective or coherent operations, dissipative relaxations, etc., have only crudely and inadequately been addressed to date. Questions such as holistic system design, self-correcting or checking architecture, the interaction of the ultrasmall structure array with the macroscopic world, packaging, etc., are also important but are downstream problems to be addressed only after the fundamental questions are answered. While architecture is currently of very fundamental importance, expectations of new structural opportunities that may arise from this program are somewhat premature; at present it would not be wise to emphasize architectural questions unless structural and transport problems have been resolved. Further programs addressing architecture on a larger-dimensional scale are underway in other DoD efforts; for example, VHSIC.

As in the other electronic areas, USER will have to compete for scarce manpower resources. At present, few U.S. scientists seem to work in the field of ultrasmall electronics research and it will be necessary to generate a national momentum to attract good scientists to this endeavor. Preliminary steps in this direction have already been taken: invited and contributed papers were given on this subject at well-attended national conferences of relevant professional societies. In the future, widely publicized, open workshops and topical conferences will communicate the need for this research in USER. However, it is recognized that the interest and cooperation of good scientists for this program will depend mainly on the *reliable availability of financial support.* The long-range nature of the USER program is being emphasized in that regard.

Finally, a note on funding. A special National Research Council panel [3] examined the funding question and strongly recommended a program of at least $10 million/year which would include USER, lithographic, and fabrication problems. For the USER program an approximate initial funding level in the neighborhood of $6 million is expected in fiscal 1981, and growth to a level of $12 million is expected by 1984. These funds will be spread over contract research programs of the three research offices in the military departments: Army Research Office (ARO), Office of Naval Research (ONR), and the Air Force Office of Scientific Research (AFOSR). In addition, the existing expertise in the prestigious Joint Services Electronics Program (JSEP) will be utilized as well as block-funded programs similar to the Selected Research Opportunities Program of ONR to establish a broad program of support of ultrasmall electronics research. Also, complementary programs at several DoD laboratories will be started. For example, both the Naval Research Laboratory and the Army's Electronics Technology Development Laboratory at Fort Monmouth are actively planning to pursue research programs in this area. Since USER will be strongly interdisciplinary, involving electrical engineers, physicists, metallurgists, and mathematicians, it will be important to optimize the synergism among these groups both in-house and on-contract. Again, JSEP will be used as a comparative example since it draws these groups together several times a year.

The DoD had a goal in the late 1950s of achieving molecular electronics. That goal was not reached, but the electronics era, spearheaded by the development of the integrated circuit, was born. Now, 20 years later, the DoD still has a goal of reaching molecular electronics. With USER, we believe that goal is achievable; at worst we might stumble on another innovation like the IC.

B. The DoD VHSIC Program—A Government Perspective

John O. Dimmock[2]

1. Background and Motivation

Although it had been developing over several years, the clear need on the part of Department of Defense to initiate an increased emphasis in integrated circuit technology became apparent in the latter half of 1977. Following the major role played by the DoD in supporting the integrated circuit industry in the early 1960s, the department deemphasized support of this technology, being convinced that industry would pursue the development without further major DoD support. This expectation has turned out to be correct, it is suspected, far beyond the projections of the original analysis. Inspired largely by a rapidly growing commercial market, integrated circuit technology has grown exponentially over the last 20 years. During the same time period the fractional share of the market commanded by the DoD dropped from about 70% in 1965 to about 7% at the close of the 1970s, and this share is still declining today. One should realize that during this period the DoD dollar volume in the market actually increased but at a miniscule rate compared to commercial demands.

These events have had a substantial impact on the utilization of current integrated circuit technology by the DoD. Substantial differences exist between DoD requirements and commercial requirements for ICs. Although many commercial applications involve somewhat hostile environments, mostly mechanical and thermal stress, none are as severe as those envisioned by the DoD. Although many commercial applications

---

[2] Physics Division, Office of Naval Research, Washington, D.C.

put stress on reliability, as in vehicle and equipment control, the DoD demands are more stringent. Although many commercial applications require that circuits can be tested and faults diagnosed, DoD applications stress this requirement more severely. Finally, DoD requirements focus more on high-speed signal processing forced by extremely high input data rates from a variety of sensor and communications systems. This forces the military to consider faster clock rates than are generally required by commercial systems. Thus commercial ICs fall short of DoD requirements largely in terms of reliability, testability, immunity to environmental stress, and speed. As a result, ICs for DoD systems either need to be custom selected and tested or custom designed and fabricated. This leads to higher cost but also, and in many ways more significantly, to a substantial lag of several years between the commercial availability of a circuit and the availability of its military counterpart. With the IC technology advancing so rapidly, this means that military ICs are substantially behind the commercial state of the art in processing capability.

This unfortunate situation has been developing gradually and, although it is significant, it probably would not have been sufficient to motivate a major DoD initiative at this time. Another development was necessary. This was the determination that our lead in military electronics IC technology, a lead that had been considered substantial, had been greatly eroded. In contrast to the present commercially dominated situation in the U.S., the IC market in the Soviet Union appears to be dominated by the military, like the U.S. market was in the early and mid 1960s. Largely because of this, Soviet military IC technology has substantially closed the technology gap. In a balance of power in which the U.S. presumes to trade off superior technology against greater numbers of weapons, this discovery was and is highly motivating. Integrated circuits are a recognized force multiplier for the DoD and it is viewed as critical that steps be taken to prevent the further erosion of America's current technological lead.

It was largely this motivation that gave birth to the very high speed integrated circuits (VHSIC) program in 1978 [4–6]. Although a precise date of initiation of the program is difficult to establish, perhaps July 19, 1978, will serve as well as any other. On this date the Under Secretary of Defense for Research and Engineering, William J. Perry, sent to the Assistant Secretaries of the Army, Navy, and Air Force a memorandum with the subject "Very High Speed Integrated Circuits." The following is a quotation from that memo:

I believe that a major new DoD initiative is essential to develop a new generation of very high speed ICs (VHSIs) suitable for rapid deployment in military systems. The goal of this initiative will be to provide these advanced ICs for military use at least five years earlier than present commercial trends indicate. These VHSIs will operate with a processing speed 50 to 100 times greater than present ICs so that a single IC could replace many present ICs. They will provide a 10-fold reduction in size, weight, power consumption, failure rate and life cycle cost of present military computer systems. Additionally, their high processing speed can provide new and important capabilities for many of our military systems such as cruise missiles, satellites, avionics, radar, undersea surveillance, electronic warfare, SIGINT [signal intercept] and C^3 systems. William J. Perry

Some indication of the managerial and technical complexities of the ensuing VHSIC program can be obtained by noting that the initial phase of the main program, referred to as the Program Definition phase, Phase 0, did not start until 1980 and Phase I did not start until 1981. Nevertheless considerable interactive planning has transpired among the three services, the Office of Under Secretary of Defense for Research and Engineering, industrial interest (both pro and con), universities, and, of course, the Congress. It is not the purpose of this synopsis to detail the technical, managerial, and political interactions that brought the program to its present formulation; in fact at the time of this writing the VHSIC managerial structure is still not fully established. It is enough to say that these interactions have been, in general, constructive and the program is much better defined now than it was in 1978. As planned at present, the VHSIC program will be directed by the Office of the Deputy Under Secretary for Defense for Research and Engineering (Research and Advanced Technology) with individual contracts managed by the three services. A total of $200 million is planned for the program to be spent over a seven-year period.

2. Program Description

The VHSIC program is divided conceptually into four phases, 0, I, II, and III. Phases 0, I, and II will be carried out consecutively whereas Phase III will be carried out in parallel with the other three phases. Figure 1 shows the anticipated time sequencing of these phases. Phase 0 contracts were signed on March 7, 1980, and the program is expected to continue into 1986. Each phase of the program has distinct goals although there is considerable cross fertilization intended between phases. In particular, although Phase III is designed to provide high-technology support for submicron development, it is expected that early results from this phase will benefit Phases Ia and IIa as well.

The overall objective of the VHSIC program is to develop militarized,

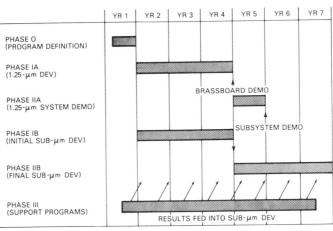

Fig. 1.   VHSIC program schedule.

advanced ICs to be introduced into future military systems in a timely and affordable manner with a minimum of customization. The expected benefits from this include advanced military capability, high return on investment including life cycle costs, and an increase in our lead in military ICs by several years. The program expects to advance vastly not only IC technology as measured in terms of computational speed and efficiency and to reduce system size, weight, and power requirements, but also to achieve this advance directly in militarized circuits with high reliability, ease of testability and diagnosis, and high tolerance to military stress environments. Each phase of the VHSIC program is intended to play a specific role in reaching these objectives.

Phase 0 is intended primarily to be a study phase to define the work necessary, the detailed approach, and the plans for achieving the ultimate objectives of the VHSIC program. This program analysis is expected to proceed in a top-down fashion starting with the selection and analysis of at least three projected military systems or subsystems to determine their signal and data processing requirements and to identify broadly applicable VHSIC chips or modules required. From this a common set of VHSIC "building blocks" will be defined, out of which a maximum of military, high-performance, electronic systems could be constructed with a minimum of customization. It is not clear at this writing at what level of integration this modularization will occur. In some sense a VHSIC chip already will be an integrated system and it is envisioned that many systems will consist of only a few chips.

With increasing integration has come a strong trend toward customization as each chip becomes a system in itself. The microprocessor is probably the only example in which this is not the case. However, microprocessors and microprocessor-based architectures are likely to be too slow and inefficient to meet VHSIC goals. In fact, in Phase 0 architectures and design approaches will be selected and investigated to implement VHSIC chips with 1.25-μm and submicrometer minimum feature size. For the 1.25-μm feature sizes (in preparation for Phase Ia), devices will operate at a minimum clock rate of 25 MHz and will provide a functional throughput of a minimum of 5×10^{11} as expressed in the system figure of merit clock rate times gates per square centimeter. The throughput will be approximately 10^{13} clock rate times gates/cm^2 for the submicron features (in preparation for Phase Ib). In addition, chip technology and processing techniques will be specified. This will include device technologies, interconnects, and metallizations appropriate for both the 1.25-μm and submicrometer feature sizes as well as chip fabrication and manufacturing techniques. The studies also will include packaging trade-offs, computer-aided design (CAD) techniques, and lithographic requirements and techniques all necessary to achieve the overall VHSIC program goals. In addition, susceptibility to thermal stress over the full military temperature range from -55 to $+125°C$, to man-survivable transient and continuous radiation doses, to electromagnetic interference, and to other environmental factors met in military applications will be considered.

One of the more difficult aspects of the program is the requirement to define and describe a viable and effective procedure for making VHSIC components available to all other DoD contractors and government laboratories in a timely and affordable manner. This includes the agreement to license and assist government-designated parties to use contract products for government purposes including technical data and computer software. Finally, under the Phase 0 effort, plans are to be described for rapidly introducing VHSICs into DoD systems.

The main intent of the Phase 0 program is to enable contractors to study and analyze the VHSIC program requirements sufficiently to determine what problems are likely to be met and what approaches are likely to be the most successful. The result is expected to be proposals for Phase Ia (and/or Ib).

As indicated in Fig. 1, Phase I is subdivided into two parallel efforts. Phase Ia is directed to the development of complete electronic brassboard subsystems within about three years of the Phase I start. These brassboards may consist of several VHSIC chips that are "building block" modules. The modules will have a minimum level of customization, they will use 1.25-μm feature size technology, they will have an equivalent gate clock frequency product exceeding 5×10^{11} gate Hz/cm^2 and a minimum clock rate of 25 MHz. It is anticipated that a pilot line production capability will be established for this technology. Minimum requirements of reliability, testability, and environmental immunity will be demanded.

Phase Ib will consist of initial efforts to extend IC technology to submicron feature sizes and corresponding circuit complexities. This will include high-resolution lithography and replication techniques, submicron device design and modeling, substrate and epitaxial growth improvements and analysis, metallization reliability and interconnect analysis, appropriate CAD techniques, and architecture and systems considerations. A feature of both Phases Ia and Ib will be the development of built-in, on-chip, testing technology including design for testability. Specific requirements for reliability and testability should be developed in the Phase 0 efforts.

Similarly, Phase II is divided into two parallel programs, Phase IIa, which will provide subsystem demonstrations based on Phase Ia brassboards, and Phase IIb, which continues the Phase Ib submicron development effort. Phases Ib and IIb are focused at developing all aspects of IC technology necessary to cross the so-called 1-μm barrier, which is considered by many to be the practical limit of conventional optical lithography and fabrication techniques. The end goal of these efforts is the development of production capability for such advanced chips. This includes not only lithographic and fabrication, but also design, architecture, software, and testing technologies. Subsystem and system demonstrations of Phase IIb technology are expected at the end of the program or shortly thereafter through extensions or other service funding.

The Phase III VHSIC support program runs in parallel with the main program efforts. Whereas the Phase I and II program contracts are expected to be large, vertically integrated efforts with each contractor covering all aspects of VHSIC development, Phase III will consist of many smaller, shorter-term efforts in key technology areas designed to feed into the main program. VHSIC III support programs are considered essential to develop a solid technology base and to stimulate innovation and provide sources for specific design, manufacturing, and test equipment. Efforts will focus on high-resolution lithographic equipment and processing technology; advanced architecture and design concepts for reducing custom fabrication, increasing chip utilization, and improving system reliability through fault tolerance and system testability through on-chip testing; advanced CAD techniques; improved packaging for electrical, thermal, and mechanical stress tolerance; improved silicon materials and fabrication processes; analytical methods for determination of substrate and fabrication-induced defects at the submicron level; methods for improving radiation, thermal, and mechanical stress tolerance; establishment of design standards and interface requirements; new device, gate, and circuit structures; techniques for documentation; and methods for improved and simplified utilization and testing.

In essence, all of these efforts can, and most should, be carried out under the vertically integrated efforts in the main program. Phase III efforts are intended to reduce the risk in the overall program through many innovative efforts on specific problems of concern.

3. Concerns and Conclusions

From many points of view VHSIC is a high-risk program, although from a DoD perspective, as previously indicated, the undertaking of such a program is critical at this time. There appears to be little doubt that the VHSIC program will accelerate the availability of advanced, militarized ICs. It also appears quite probable that many of the ambitious goals of the program will be accomplished in at least close to the projected time frame. It also appears probable that Phases Ia and IIa brassboard and subsystem demonstrations of 1.25-μm technology ICs can be accomplished in approximately the time projected. However, questions remain at this time as to the degree of standardization that can be usefully achieved or, in other words, how much customization will be required for specific applications. This will impact cost and ultimate utilization as well as reliability and testability. At this time, questions still remain regarding the meeting of full military environmental requirements of temperature range, electromagnetic immunity, and radiation tolerance. Perhaps the major questions to be addressed in the case of the 1.25-μm technology by both contractors and government are concerned with technology transferability within the U.S., prevention of this transfer to adversaries, and development of methods for rapidly incorporating the VHSICS into DoD systems. This is not meant to downplay the other technological risks in this part of the program but they appear to be less severe than those just noted.

All of these concerns are even more important in the case of the submicron technology portion of the program. In addition, because of the greater uncertainty of this technology, there are numerous additional technological questions. What new substrate- and fabrication-related defects and defect-generation mechanisms are important at submicron dimensions? What effect will these have on yield? What new limits in device performance are expected and what modifications must be made in device models for these geometries? What new device failure mechanisms are expected? Will device structure, contacts, metallizations, and interconnects be sufficiently stable under stress to maintain normative operation for required low failure rates? What limits will be imposed by natural radiation-induced failures? What major chip architecture design changes will be necessary to realize the required functional performance while achieving yield, testability, and failure rate goals of the program? Will the yield and reliability of interconnect crossovers limit the architecture or the degree of integration? Will the modular approach still be valid at this higher level of integration where each chip is a system? How much customization will be required and what will this do to reliability, testability, and cost? How can these chips be incorporated into DoD systems?

The last question is probably the most critical in the paragraph. Major system redesigns are likely to be required to utilize fully the expected output of the VHSIC program. If the large delays presently experienced in introducing IC technologies into military systems are to be substantially reduced, not only does the VHSIC technology need to meet military specifications, but advanced systems designers need to be thoroughly aware of this technology and convinced that the VHSIC chips will function according to specifications and be available at an acceptable cost. Thus there needs to be a continuous interaction among the DoD, industrial electronics systems engineers, and the VHSIC program. This appears to be well recognized by the program managers but will need to be maintained over the life of the program.

From the viewpoint of the DoD, the VHSIC program is a large, extremely important, even critical undertaking. It is receiving high-level attention by all three services and by the Office of the Under Secretary of Defense for Research and Engineering. It is viewed as critical at this time to maintain our lead in military electronics systems and information-processing capabilities. In an era in which the U.S. of necessity is emphasizing technological superiority rather than superiority in numbers of weapons systems, in which battlefield horizons are extending, in which projected information rates are exploding, and in which electronics technological superiority is becoming perhaps the most critical factor of battlefield success, the VHSIC program to a large degree is central to maintaining a strong defense posture.

C. The DARPA Program

In addition to VHSIC and USER programs and efforts at DoD laboratories, a major DoD-funded VLSI program has been mounted by the Defense Advanced Research Projects Agency (DARPA). Two decades ago DARPA was established to initiate funding of high-risk, high-potential-impact research programs at the forefront of defense-related technology. The agency also serves as an advisory resource to DoD administrators in research-related matters. Since DARPA views VLSI as an important element in its mission, it has developed a funding program whose planned level in the early 1980s is in the $15–$20 million/yr range.

The DARPA program has four basic elements: (1) basic materials, device physics, and processing research; (2) submicron digital technology; (3) architecture and design for VLSI; and (4) technological applications. The larger initial effort in the program concentrates in the first two areas, which include such long-range topics as gallium arsenide devices, graphoepitaxy, and focused ion-beam research as well as feasibility demonstrations of submicron systems. The third area is concerned with developing innovative architecture and design techniques for chips that contain more than a million gates and focuses on fast turnaround design methods. An interesting experiment to be explored in this context is the remote fabrication of VLSI systems at industrial sites using designs supplied via communication channels like the ARPANET system. The fourth element of the VLSI program is concerned with applications of VLSI systems and, eventually, is expected to impact virtually the entire DARPA structure. The overall DARPA program is directed toward developing the relevant potentialities of VLSI systems by the mid to late 1980s.

III. SCIENCE AGENCY PROGRAMS

The science and technology agencies of the federal government approach VLSI in diverse ways depending on their missions and objectives. Three examples of agency activities presented here should illustrate the scope of this diversity.

A. The Department of Commerce and the Role of Government Standards

W. Murray Bullis and Robert I. Scace[3]

1. Past Programs

Metrology has been a significant factor in the advancement of the semiconductor device industry. Understanding and control of both dimensional characteristics and materials properties have been essential in the development, design, and production of transistors and integrated circuits since their invention. The trend to VLSI increases the metrological requirements associated with each of these areas [7].

The National Bureau of Standards (NBS), as the national research laboratory for measurements and physical standards, has been concerned with measurements for semiconductor materials and devices since about

---

[3] Electron Devices Division, National Bureau of Standards, Washington, D.C.

1960 [8]. From the start, this activity has been very selective; research projects have been chosen and carried out on the basis of close interactions with the producing industry, its suppliers, and users in other Federal agencies and in the private sector. Early projects on silicon resistivity and transistor second-breakdown measurements have expanded to become the Semiconductor Technology program, which covers a broad spectrum of research activity on techniques for characterizing electronic materials, including silicon and other semiconductors, insulator films, metal interconnects, process chemicals, and resists, and techniques for measuring pattern dimensions, selected electrical and thermal properties of devices, interconnection and die bond quality, package integrity, and other critical quantities. The primary outputs of this program include improved measurement methods, prototype measuring instruments, interpretive theory, and supporting data. It is important to emphasize that it does not include development of new device designs, new processing or assembly techniques, or new manufacturing equipment, except as these may be ancillary to the primary outputs.

This activity is carried out at the Washington complex of the NBS in the Electron Devices Division, which is a part of the Center for Electronics and Electrical Engineering, one of the bureau's thirteen scientific and technical centers. Financial support for the work has come both from direct appropriations and from funds transferred by other Federal agencies, principally the Departments of Defense and Energy and the National Aeronautics and Space Administration. Most of the bureau's research is conducted in-house; in the present program a significant fraction of the research that was funded by DARPA was carried out by outside contractors during the late 1970s, but major contract activity is not expected to continue into the future.

Among the early projects with the greatest impact on the semiconductor industry are those concerned with the four-probe method for measuring resistivity of silicon slices [9], test methods for wire bonding [10–12], and thermal resistance measurements on power transistors [13]. Research on each of these topics was undertaken in response to defined industrial needs as articulated by an industrial standards committee or by a concerned Federal agency. Identification of priorities for NBS work in this way is particularly important because NBS does not establish specifications, is not a major procuring agency, and has no regulatory authority. The results of the work were implemented in many different ways which are typical of the diverse relationships between the NBS program and the industry.

In the case of the resistivity work, which was undertaken in response to a request from Committee F-1 on Electronics of the American Society for Testing and Materials (ASTM), results were implemented in three phases.

Early interim results, such as geometric correction factors, data on the temperature coefficient of resistivity, and specimen preparation techniques, were reported to the committee and in the literature as they were obtained. In this way interested parties were able to incorporate the findings in their routine measurement procedures for both materials acceptance and process control before the project was completed. As the work neared completion, the results were incorporated into an ASTM standard [14] which became the referee method for the industry and also into four other related ASTM standards. Finally, several types of sets of silicon slices were certified as to their resistivity as measured in accordance with the ASTM method [14] and issued as standard reference materials (SRMs) by NBS [15]. These SRMs continue to be useful reference standards; although the industry is presently shifting from the four-probe method to a contactless technique [16] for routine resistivity measurements, the newer method requires the use of standards to provide calibration at several values of resistivity.

The wire bond work was undertaken in response to the needs of a military agency. The initial output was a new method for monitoring the ultrasonic bonding process [10]. Use of this method together with a novel scanning electron microscope examination technique [17] led to greatly increased understanding of the process, to suggestions for improvements in bonding machine design (many of which were incorporated by equipment makers in retrofits or next-generation machines and thus introduced painlessly into the device manufacturing process), to a sound basis for the nondestructive pull test [18], and to substantial improvements in productivity and device and system reliability. In this instance, the results were incorporated first on captive lines operated in several companies on behalf of a major military system; they then diffused through the industry and were eventually embodied in ASTM standards [19,20] and in the bond test methods of the military standards for discrete devices [21] and integrated circuits [22]. The interlaboratory test to evaluate the destructive pull test [20] identified and helped solve a variety of problems including a potential deficiency in a widely used automatic pull tester [23].

The thermal resistance work was an outgrowth of the early work on second breakdown in transistors. The project was carried out in close collaboration with various committees of the Joint Electron Devices Engineering Councils (JEDEC). The work led to identification of a preferred electrical test method and its range of application which was issued as an Electronic Industries Association Recommended Standard [24] and incorporated in a military standard [25]. The results of this work formed the basis for a commercial thermal resistance test instrument and eventually led to a new method for determining the safe operating area of power transistors [26].

2. *Future Directions*

Some of the more recent projects in the program related directly to some of the metrological requirements of VLSI. One critical requirement relates to dimensional metrology of integrated circuit pattern features in the 0.5–10-μm range on photomasks [27] and wafers [28]. This work was undertaken following an extensive industry survey [29] that established the need for linewidth measurements in this regime. Procedures have been developed for making linewidth measurements on photomasks; SRMs are being made available to the industry; and ASTM standards are under development. These procedures have also been extended to patterns in very thin oxide layers on silicon; extension to patterns in technologically more interesting layers greater than 0.2 μm thick is underway. In addition, the techniques are being transferred to the industry in a series of training seminars featuring hands-on experience with linewidth measuring systems loaned for the purpose by instrument manufacturers. From time to time, multilaboratory tests are conducted to assess the state of the measurement art in this field.

Microelectronic test structures are devicelike elements used to measure selected critical material and process parameters by means of high-speed electrical tests [30]. They are particularly important in connection with design rule verification and in establishing, verifying, and controlling wafer fabrication processes. Work in this area was undertaken following both extensive informal discussions with industry representatives and conduct of a workshop [31] held to provide a more formal channel for discussions on this topic. A wide variety of structures have been investigated including a cross-bridge array for evaluation of pattern generators and step-and-repeat cameras, a production-compatible potentiometric electrical alignment test structure, which has been shown to be sensitive to misalignment of 0.1 μm, and both serial and random access structures to test for random fault distributions. In addition, test structures for use in measuring material properties with conventional automatic wafer probers and d–c instrumentation are also being studied. Among the more advanced of these are the four-terminal enhancement-mode MOSFET, with which dopant profiles can be measured [32], and the gated diode test structure with integral transistor switch and source-follower MOSFET electrometer, which facilitates the measurement of small leakage currents and the determination of generation lifetime and surface recombination velocity [33].

Both the dimensional metrology and microelectronic test structure efforts are being extended in connection with DoD's program to develop very high speed integrated circuits. In addition, the Commerce Department soon plans to undertake an initiative on basic measurements for

VLSI. This initiative is intended to provide the same types of technical outputs as the ongoing Semiconductor Technology Program, extended to accommodate the denser circuits, larger chips, and new processing techniques associated with VLSI. The augmented program will address the following six critical areas: (1) silicon characterization; (2) interface characterization; (3) process control metrology; (4) microelectronic test structures; (5) package evaluation; and (6) testing.

Although rather detailed preliminary plans have been developed, it is essential to emphasize that the program will respond to industry requirements as they develop. It is also important to realize that, as has been done in the NBS Semiconductor Technology program heretofore, many of these activities involve much more than simply finding a way to make a measurement. Fundamental work to develop a clear understanding of the basis for the measurement and its range of utility is often required as well. Some of the key issues to be expected within each area include the following:

a. Silicon Characterization. Within the area of silicon characterization, some key issues are the methods for profiling the distribution of dopant impurities, especially in shallow, low-dose ion implantations; the methods for characterizing nondopant or deep-level impurities and the resolution of differences among results obtained by various methods; the identification of appropriate parameters to be specified for substrate wafers; and the techniques for characterizing wafer surface distortion.

b. Interface Characterization. Within the area of interface characterization, some key issues are the methods for measuring neutral trap densities and distributions near the silicon–silicon dioxide interface; the techniques for characterizing the structural and electrical properties of this and other device interfaces; the techniques for evaluating multilayer metal interconnect systems; and the methods for measuring contact resistance at the interconnect-semiconductor interface.

c. Process Control Metrology. Within the area of process control metrology, work will be directed toward determining the methods for measuring line-width and edge quality in small features with large-aspect ratio (height to width); the measures for sensitometry and other attributes of resists; and the metrology associated with control of processes such as plasma etching, ion implantation, laser annealing, and ion milling.

d. Microelectronic Test Structures. Within the area of microelectronic test structures, the following issues are important: the scaling to VLSI dimensions of test structures for measuring material characteristics, pattern definition and registration, and random fault densities and destributions; the test structures with integral signal processing circuitry; the correlations between test structure characteristics and device or circuit properties; and the dynamic and reliability test structures.

e. Package Evaluation. Within the area of package evaluation, some key issues are the methods for measuring device temperature and package heat transfer characteristics; the methods for evaluating mechanical integrity; and the tests for hermeticity and internal moisture levels.

f. Testing. Within the testing field, emphasis will be placed on the development of models for fault mechanisms; this portion of the effort is expected to link with NBS programs being developed to address broader aspects of testing complex integrated circuit chips and systems.

This program continues into the VLSI era the role of the federal government in the development of procedures and specifications to improve device performance, product yields, and reliability assurance begun in the earliest days of the transistor. For example, during the early days of the development of the integrated circuit, the Rome Air Development Center introduced MIL-STD-883, Test Methods and Procedures for Microelectronics, and MIL-M-38510, General Specification for Microcircuits, which became accepted as the worldwide standard for quality control of small- and medium-scale integrated circuits. These two military standards have been developed further over the intervening years and now provide sampling plans and test techniques for control of the quality of integrated circuits at the highest levels. The line certification criteria for production and acceptance of MOS semiconductor devices, developed by the NASA Marshall Space Flight Center, were instrumental in improving the manufacturing methods for these devices through improved in-process measurement and control, as was the standard PMOS test cell family introduced in the late 1960s by another DoD agency. The goal of the expanded NBS program is to provide the more sophisticated metrology, including the underlying knowledge and supporting data, essential for VLSI.

B. The NSF Role in VSLI

Although the National Science Foundation has had a long interest in supporting basic aspects of semiconductor electronics, solid-state physics, and other sciences that underlie VLSI, the foundation's role in direct support of VLSI began with the establishment of the National Research and Resource Facility for Submicron Structures located at Cornell University [34]. The support continues under a more broadly based microstructures program. Within the traditional NSF realm of university-based research, the foundation's efforts to develop activity in VLSI have probably been significant. Through a series of workshops [35–37] and a competition to determine the host institution for the national facility, considerable attention was focused on submicron microfabrication and on associated electronics-related research issues. These efforts raised the consciousness of university researchers and administrators and led to some program initiation. Of course, the present high level of university interest in VLSI owes at least equally to establishment of VHSIC and other DoD programs and their promise of substantial continued financial support. NSF's participation in submicron-related activities has also benefited the foundation and its scientific constituency as it has demonstrated to the Congress the long-term relevance of its activities to industrial development.

1. The Submicron Facility

Advancement of VLSI was not the original NSF objective in establishing the Cornell facility nor is it likely to be a predominant theme in future programs. The foundation is an independent agency of the federal government that was established to advance scientific progress in the United States over the broad range of scientific activities. Foundation policy is to act in concert with mission-oriented agencies of the government or with industry in support of research and not to compete with them or to duplicate efforts. When the submicron program was begun, there was a high degree of sensitivity to the predominant role that industry would play in electronics research and a strong attempt to minimize the VLSI implications of a submicron program. That sensitivity has been diminished to a large extent because of an increasing acceptance of the importance of a university research and training role in VLSI.

The origins of the NSF program stem from a response to a 1974 request from Engineering Division management that the staff explore research opportunities involving large grants rather than traditional single-investigator projects. Many suggestions were made but only the submicron program survived the rigors of foundation review and was funded in the 1970s. The submicron program was suggested because of the author's own prior involvement in integrated optics research and his consequent awareness of industrial efforts in the microfabrication area, the scope of skills and facilities needed to provide a significant laboratory, and the scientific and intellectual potentialities of the field. A number of papers, including a 1973 talk by Henry Smith [38], contributed to this view. The establishment of a submicron research and resource facility was possible because the identification of a small size frontier carried with it enough intellectual promise to gain the attention of NSF management accustomed to the exotic stimulation of elementary particles and black holes. Rediscovery of the interest of Richard Feynman in small structures [39] may well have been helpful in that regard. The establishment of a resource fabrication facility also appealed to NSF because it could potentially aid a broad cross section of researchers including those at institutions not likely to develop facilities of their own.

The relationship of submicron microfabrication to VLSI was a complex issue. The reality of a close relationship meant that the risk involved in starting a major program that electrical engineers might not find interesting to pursue was minimal since electronics is so central to electrical engineering. From the position of the National Science Board, which is the governing body of NSF and is responsible for approving major projects, the relationship to VLSI was a concern since, from their perspective, industry could be expected to do the research needed for the advancement of VLSI. Some DoD personnel apparently shared this point of view and they declined to participate in establishment of a university-based submicron facility. The relationship of submicron techniques to engineering and science concerns outside of VLSI was emphasized in every presentation to the board, although never with clear-cut success.

The board also required substantial community sampling before it would approve funding of a facility. A series of workshops was held in which the views of several hundred university and industry research personnel were obtained. There was some minimal opposition to the program, most noticeably from some Bell Laboratories personnel who later became active and helpful participants in the program in a manner reminiscent of an earlier attitude toward initial establishment of the foundation itself [40]. In October 1976, the board passed the resolution approving a solicitation of proposals. A total of eighteen proposals were received from diverse departments of electrical engineering and they were reviewed by a qualified eighteen-member panel that had a broad university–industry–government representation. The panel ranked the Cornell proposal first, but, in general accordance with the review guidelines, they recommended four institutions for a final site visit by NSF and consulting staff. In the end a strong Cornell faculty and administration commitment to the facility coupled with demonstration of a substantial university program in the basic science areas relevant to microfabrication won the facility for Cornell. In November 1977 the board passed a resolution authorizing a grant to Cornell University to establish the facility.

2. The Microstructures Program

The extensive three-year effort to launch the submicron facility was followed by a period of further support for small structure research. James Krumhansl, then Assistant Director of the Foundation for Mathematical and Physical Sciences and Engineering, was apparently struck by the broad relevance of the submicron area to the fields under his purview. A significant example of that relevance was the work of Ruoff [41] in the area of high-pressure phase changes in materials. To perform his research, Ruoff needed a way to measure the resistance of materials compressed under a diamond point of submicron radius of curvature and one way to do it was through fabrication of submicron interdigital structures. Other submicron-related research ranging from heterogeneous catalysis to x-ray imaging was identified [42] and in fiscal year 1979 a special allocation of $1.8 million in foundation reserve funds was prescribed for support of microstructures research. The funds were distributed over a number of foundation activities including electrical science, solid-state physics, chemistry, ceramics, mathematics, and computer science and were used to support a broad range of proposals that had been previously reviewed for general scientific merit. About a third of the funds were used directly to increase support for electronics-related research.

The federal government's interest in basic research in small structures has resulted in special allocations to the foundation for support of microstructures activities. The allocations began in fiscal year 1980 with $2 million with increases the next year of an additional $4 million. The total microstructures program will be just under $15 million/year by that time. The relatively wide distribution of these funds among the research areas supported by the foundation, however, will limit their visible impact on VLSI.

There are a variety of forces that will shape future development of NSF support for VLSI-related research. A dominant one is the view that support for basic research in general will aid the advancement of applications areas like VLSI, albeit in unpredictable ways. This is a supportable view and, of course, it is also one that helps the various groups in NSF to share in the microstructures funding. Since microstructure research contacts diverse areas like photochemistry, surface physics, computer-aided design, crystal growth, biomedical instrumentation, and optical communications, it provides a useful vehicle for broad funding activities. The broad funding approach is likely to characterize future NSF activities in microstructures. Among the implications is that new block-funded ventures like the submicron facility are likely only if the existing facility proves truly invaluable in advancing multidisciplinary interests.

Another force affecting the future of the NSF microstructures program is the development of DoD programs in VLSI. The expected growth in these programs, particularly the long-range, basic research aspects, will impact the perceived need as well as the incentive for support of microelectronics by NSF, and the support, therefore, is likely to grow modestly rather than to increase in a significant way. On the other hand, VLSI research is likely to replace aspects of other electronics and computer-related research so that there is a fertile funding ground of some tens of millions within NSF in which the research can take root. As is customary for NSF, funded programs will attempt to emphasize basic research that is farsighted and imaginative.

C. The NASA Role in VLSI

James A. Hutchby[4]

The National Aeronautics and Space Administration's primary mission centers on four activities: (1) development of space for the benefit of humanity; (2) exploration of space for increased knowledge and improved understanding of our universe; (3) development of new energy-efficient and low-noise aeronautics technology; and (4) development of the space transportation system—the shuttle.

Each of these pursuits use and depend on new electronics technology to perform its function. For example, the first activity encompasses a host of earth-oriented themes and projects. These include global surveys of crop production, water availability, land use and environment assessment, prediction of weather on a large scale, stratospheric changes and effects, and transfer of information through various communications systems. These earth-oriented space-flight experiments are all required to measure an increasing number of resolution elements or pixels of decreasing size and area. Therefore, the next twenty years will see a steady and rapid growth in the type and amount of data collected in space and returned to Earth, combined with the necessity for acquiring, processing, and disseminating this information at low cost.

Several advanced electronic technologies will be required to support these new systems. These include high data rate, monolithically integrated, infrared sensor chips with some on-chip signal processing capability; on-board-spacecraft data storage, processing, and compression systems; and high-speed data communication systems.

NASA has separated the various electronics technology advancements required to pursue its four primary missions into two categories; those that will be primarily developed by private industry and will not require significant federal support; and those that are uniquely required by NASA's mission and will therefore require substantial Federal funding. Microelectronics, including VLSI, fall into this first category; yet the es-

---

[4] Energy and Environmental Research Division, Research Triangle Institute, Washington, D.C.

tablishment of long-term reliability of microelectronics in the space environment will require federal support. The second category also includes development of the monolithic, infrared imaging sensors and signal processing devices (which require fine-line lithography, dry processing, ion implantation, and other VLSI-related technologies) and high-density, magnetic-bubble, solid-state data recorders.

Accomplishment of NASA's missions, therefore, will be dependent on industrial and Department of Defense development of LSI and VLSI chip and integrated systems technologies. Once available, however, NASA will exploit their potential in various ways. These include data processing and control systems, integrated spacecraft data communications systems, and ultra-high density data storage systems. The obvious attraction to the VLSI technology is its high functional capability per unit cost, size, weight, and power expended.

The implementation of space-borne data processing and control functions using VLSI technology will provide dedicated computers applied to individual functions on board space vehicles. Initially, these microprocessors will operate independently to simply software concepts, will permit interaction of computer elements at higher levels, and will provide load-sharing and fault-tolerant operation.

In the area of communications, development of low-cost methods of arranging integrated dipole elements for achieving required aperature size may be considered in light of advances in VLSI. Not only may this become a cost-competitive approach to conventional reflector antenna design, but it will also provide for multibeam operation and electronic beam shaping and steering. In addition to the need for large-aperture antennas, special designs are required at L, S, X, and K bands for radio sensing and satellite communications. These applications will emphasize special coverage patterns, sidelobe control, and multiple frequency and polarization operation. Lens antenna technology and small-element array technology will be used to a large degree to provide these classes of antennas.

VLSI technology also may be exploited to develop compact, integrated communication systems composed of receiver amplifiers and transmitter power amplifier elements connected to their respective antenna dipole array elements. The major advantages of this integrated design approach are reduced cost, increased reliability, and the development of flexible electronically controlled antennas. Using microprocessors, the total system would also include data processing and attitude control functions.

Finally, development of complex, high data volume, real-time digital processors may impact a number of information transfer applications such as radar imaging, random access satellite communications, and detection of interstellar microwave signals in a search for extraterrestrial intelligent life.

IV. POLICY ISSUES

As VLSI systems become part of the mainstream of business and defense activities, government interest in VLSI will broaden. The nature of this broadening will be illustrated by the following discussions of policy issues.

A. Government Antitrust Policy and VLSI

Myles G. Boylan[5]

Through legislating and enforcing its antitrust policy, the government is committed to preventing the monopolization of markets, because successful monopolization frequently leads to excessively high prices. Monopolization can be achieved in a variety of ways. The act of monopolization requires that either a single dominant firm emerge or that a group of firms form some sort of cartel with the goal of sharing business, determining prices jointly, and avoiding competition. Historically, single-firm monopolization has been relatively easy to prove in court whereas cartel behavior has proven to be considerably more difficult to establish. Occasionally, spectacular abuses, such as the electrical equipment manufacturer's conspiracy of the 1950s, are prosecuted successfully.

In regard to cartels, the Supreme Court began to realize, as far back as the 1920s, that trade associations and other joint activities could provide the means whereby firms could exchange financial information and could meet (legally) to codetermine prices and market shares. Consequently, the court gradually placed restrictions on the kinds of information trade associations could collect, publish, and discuss at meetings. In additional, joint ventures and consortia were frequently found illegal if engaged in the production or marketing (except for marketing of generic products), because of the potential for exchanging information vital to establishing a cartel. Joint ventures and consortia formed to undertake R&D programs, however, have typically been permitted as long as these group endeavors have limited objectives, are unlikely to result in the exchange of confidential financial information, and do not threaten the survival of other firms in the same industry who were not members of the group. The longer a group organization of this type is designed to last, the more likely it becomes that the interest of an antitrust agency might be aroused because of

[5] Division of Policy Research and Analysis, National Science Foundation, Washington, D.C.

the increasing likelihood that a growing number of conduits for the exchange of information among the parent firms will be institutionalized.

Two concerns related to U.S. antitrust policy have arisen in the U.S. electronics industry in regard to VLSI. At present, antitrust-based remedies have not been sought and there are no programs underway within the Federal Trade Commission (FTC) or the Antitrust Division of the Department of Justice that are designed to react to present or future problems within this industry.

The Electronic Industries Association (EIA) has given testimony before the U.S. Senate Subcommittee on International Finance (January 15, 1980) concerning the growing importance of the Japanese IC consortium in the U.S. market. This consortium captured a significant share of the U.S. market after 1975 (a recession year) as a consequence of overly cautious forecasting by U.S. firms of the growth in domestic demand for ICs and unduly optimistic forecasting of growth in demand for ICs in Japan by the Japanese consortium. This event allowed the Japanese firms to capture a large share of the U.S. market, according to the EIA. It is feared that a large Japanese industry will be tempted to practice dumping in the U.S. market in years when there is excess capacity. Dumping is defined as occurring when the average selling price of a brand name product is lower in the countries importing it than in the country exporting it, after adjusting for transportation costs, taxes, and tariffs.

According to a draft report to the U.S. National Bureau of Standards by Charles River Associates Inc. [43], U.S. firms producing integrated circuits have alleged already that Japanese firms have practiced dumping in IC markets. This practice is made easier, it is further alleged, by the closed nature of the Japanese home market to foreign competition. Thus Japanese firms can charge high domestic prices and penetrate foreign markets by means of low pricing policies, allowing them to expand their sales and take advantage of economies of scale.

United States firms would prefer federally sponsored relief from dumping in the form of tariffs or quotas, according to the draft, rather than a partial exemption from the antitrust laws. However, exemptions allowing mergers, acquisitions, and joint ventures could be requested by the EIA or other groups in the future if tariffs or quotas are not imposed.

A second source of concern stems from the Department of Defense's VHSIC program. Some private firms have asserted that this program will compete with them for skilled scientists and engineers and raise the cost of private R&D programs. It is further alleged that the commercial value of the knowledge and products developed in the VHSIC program will not adequately compensate private U.S. firms for the potential erosion in their competitive position (relative to Japanese firms) in conducting private R&D due to the higher costs. The National Academy of Sciences is engaged in a study of this potential problem. A related concern is that if only a few U.S. firms participate in the VHSIC program, they would end up as sole sources vendors for the military once the research stage has been completed. If this concern is valid, it is possible that the interest of the U.S. antitrust agencies might be attracted.

B. Manpower Policies

William A. Hetzner[6]

Direct government concern with VLSI and its potential impact on unemployment and the displacement of labor has been generally limited. The Bureau of Labor Statistics of the Department of Labor has had a program to analyze the potential labor impacts of technological innovations in particular industrial sectors. To the extent that VLSI has been and is involved in these innovations, it is included in the particular industry's technology outlook report. Of special note is such a report to be released soon on the electrical and electronic equipment industry.

In addition, it is anticipated that VLSI will be one of the technologies to be considered by the forthcoming Labor/Technology Forecasting System of the Departments of Labor and Commerce. The intent of this system is to identify significant future technologies and to assess their likely effects on manpower so that workers and management can anticipate and adjust to any adverse changes.

The decline in DoD and NASA support for microelectronic research and development in universities seems to have adversely affected the supply of trained manpower to microelectronic industry. The industry has been dependent on skilled personnel, perhaps more so than other industries, and it may be necessary for the government to find ways of assisting universities in providing the necessary manpower. A bill before the Senate (S. B. 1065) to allow a federal tax credit of 25% for grants by industry to universities for basic research may be one of the ways to increase university training and improve their facilities.

The Department of State has had some interest in manpower policies with respect to VLSI through its participation on the Conference on Information, Computer, and Communications Policy of the Organization for Economic Development and Cooperation. While potential labor displacement and unemployment from the use of VLSI in manufacturing technologies has been of more concern to European governments than the U.S., the economic and social impacts of VLSI are transnational in nature

---

[6] Division of Policy Research and Analysis, National Science Foundation, Washington, D.C.

and it may be important for the U.S. government to recognize European and, to some extent, third world concerns about VLSI. Representatives of the International Labor Organization, for example, have expressed concern that microprocessors and VLSI will alter the current economic and political relations among developed and developing countries by reducing the advantages of cheap labor.

C. Export Control

Stanley Pogrow[7]

The Office of Export Administration (OEA) in the Department of Commerce is the lead agency for accepting applications for export licenses, for making regulatory decisions, and for contacting other departments that may have to rule on a particular application. The OEA has dual responsibility for both civilian and military goods. It is the responsibility of the Office of East–West Trade in the State Department to transmit to COCOM, the multinational export control agency maintained by NATO countries (less Iceland, plus Japan), those applications that must be approved by that body.

The basic goals of export control are to: (a) prevent exports that would enhance the military potential of hostile nations; (b) promote foreign policy objectives; and (c) protect the domestic economy from an excessive drain of scarce resources. In addition to our own export licensing procedures, certain exports to specified Eastern bloc countries must be submitted to COCOM for approval. Furthermore, the U.S. requires companies in the COCOM countries to apply for a reexport license from us when they export U.S. goods to a controlled country. These reexport provisions apply even if the COCOM country has export licensing requirements as stringent as ours and even if the item is an assembled product using only a few U.S. produced components. Finally, provisions enacted in 1979 require the DoD to produce a list of militarily critical goods and technology. It is anticipated that the items on this list, when and if the list is developed, will have the most stringent controls.

Semiconductor technology and products constitute a major part of all export license requests and are critical elements in the development of equitable and effective export controls. It can be expected that, over time, export policy relative to VLSI will become a significant export control issue. State of the art semiconductor technology is presently controlled under the provisions for licensing the export of commodities. The

---

[7] Division of Policy Research and Analysis, National Science Foundation, Washington, D.C.

aspects of the industry that are controlled (to varying degrees) include: (a) the chips; (b) chip fabrication technology; and (c) end products consisting of chips, most notably, computers. The only country to which such controlled goods can be shipped without an export license (regardless of whether they are an ally or neutral) is Canada.

It is very difficult to assess what effect export regulation has had on the development of the semiconductor industry in this country to date and what effect it will have on the development of VLSI. Export controls clearly restrict the overall market potential and increase the difficulty in capturing the overseas markets that do exist. The effects to date, however, appear to be minimal for the large corporations since they can presently sell all the goods they can produce. Market size is not presently a constraint. Indeed, the primary concerns of industry have been to increase the controls on the export of chip fabrication equipment and to eliminate the inefficiency with which existing controls are administered. It appears, therefore, that export controls will play a relatively minor role in determining the rate of VLSI development as compared to the factors of the availability of venture capital, R&D funds, and the rate of Japanese technological development.

This is not to say that export policy has no effects. Exports currently account for 25% of the sales of the U.S. electronics industry and contributed a net surplus in the 1979 balance of payments of $4.6 billion. In addition, there are problems with the existing export policies, which, when combined with the rapid rate of technological progress being made by the Japanese, make it conceivable that we will lose our competitive edge in existing semiconductor technology as well as VLSI. Some of the potential problems include:

(a)   The sheer problem of administering a program with 80,000 applications per year in which the product line of a large semiconductor company can approach 100,000 rapidly changing items creates costly bottlenecks and delays. The cost of complying with licensing requirements is particularly troublesome for small companies.

(b)   Reexport controls put U.S. products and subsidiaries at a competitive disadvantage. Since the U.S. is the only country with such (redundant) requirements, foreign companies have an incentive to design their products around non-U.S. produced components. Where U.S. produced components or goods are used, there is a tendency for foreign companies to ignore these provisions. This gives them a competitive advantage over U.S. subsidiaries that must comply with U.S. reexport controls.

(c)   It is not yet known whether VLSI components, products, and manufacturing technology will be placed on the military's critical technology list. If it is and COCOM does not adopt similarly restrictive standards, then it is possible that Japanese VLSI technology and components will be subject to less stringent export controls than ours. Congress has already mandated that VHSIC be placed on the munitions control list.

(d) Industry and government officials both concede that a very large black market of undetermined size exists in the semiconductor industry for the purpose of bypassing export controls. Rumor has it that foreign companies in at least one COCOM country are key distributors in this black market. Major growth in the black market would not only put companies engaging in legitimate trade at a disadvantage, but it also would signal the folly of attempting to prevent the dissemination of broad-scale technological advances.

It is not clear what the overall impact of these problems will be on the rate of development of VLSI technology in the U.S. They do, however, suggest some necessary policy revisions in U.S. export policies and administrative procedures as well as the need to better coordinate those policies with allies.

ACKNOWLEDGMENTS

I want to express my appreciation both to the contributors, who provided informative and stimulating discussions on brief notice, and to Miss Ann Evans, who skillfully prepared the manuscript.

REFERENCES

1. Speakout, Electronic Engineering Times (January 7, 1980).
2. The President's Industrial Innovation Initiatives, Fact Sheet, Office of the White House Press Secretary (October 31, 1979).
3. Microstructure, Science, and Technology. National Research Council Report, Washington, D.C. (1979).
4. L. R. Weisberg, The new DoD initiative in integrated circuits, Technical Digest of the International Electron Devices Meeting, p. 684 (1978).
5. L. W. Sumney, The Department of Defense program in very high speed integrated circuits, "IEEE Computer Society Workshop on Microcomputer Firmware/Software Technology Trends Versus User Requirements." Johns Hopkins Applied Physics Laboratory (1979).
6. R. M. Davis, The DoD initiative in integrated circuits, *Computer,* July 74 (1979).
7. W. M. Bullis, "Semiconductor Measurement Technology," Metrology for Submicrometer Devices and Circuits. NBS Spec. Publ. 400–61. National Bureau of Standards, Washington, D.C., 1980.
8. W. M. Bullis, "Measurement Methods for the Semiconductor Device Industry—A Review of NBS Activity," NBS Tech. Note 511. National Bureau of Standards, Washington, D.C., 1969.

9. W. M. Bullis, Standard Measurements of the Resistivity of Silicon by the Four-Probe Method, NBSIR 74-496 (August 1974).
10. G. G. Harman and H. K. Kessler, "Application of Capacitor Microphones and Magnetic Pickups to the Tuning and Trouble Shooting of Microelectronic Ultrasonic Bonding Equipment," NBS Tech. Note 573. National Bureau of Standards, Washington, D.C., 1971.
11. G. G. Harman (ed.), "Semiconductor Measurement Technology," Microelectronic Ultrasonic Bonding, NBS Spec. Publ. 400-2. National Bureau of Standards, Washington, D.C., 1974.
12. J. H. Albers (ed.), "Semiconductor Measurement Technology," The Destructive Pull Test, NBS Spec. Publ. 400-18. National Bureau of Standards, Washington, D.C., 1976.
13. S. Rubin and F. F. Oettinger, "Semiconductor Measurement Technology," Thermal Resistance Measurements on Power Transistors, NBS Spec. Publ. 400-14. National Bureau of Standards, Washington, D.C., 1979.
14. Standard method for measuring resistivity of silicon slices with a collinear four-probe array, ASTM Designation F 84, "Annual Book of ASTM Standards," Part 43 (November 1979).
15. G. A. Uriano, The NBS standard reference materials program: SRMs today and tomorrow, Standardization News 7 (9), 8–13 (1979).
16. G. L. Miller, D. A. H. Robinson, and J. D. Wiley, Contactless measurement of semiconductor conductivity by radio frequency free-carrier power absorption, Rev. Sci. Instrum. 47, 799–805 (1976).
17. Reference [4], pp. 73–79.
18. G. G. Harman, A metallurgical basis for the non-destructive bond pull-test, Ann. Proc. Reliability Phys., 12th, Las Vegas, Nevada pp. 205–210 (1974).
19. Standard recommended practice for non-destructive pull testing of wire bonds, ASTM Designation F 458, "Annual Book of ASTM Standards," Part 43 (1979).
20. Standards methods for measuring pull strength of microelectronic wire bonds, ASTM Designation F 459, "Annual Book of ASTM Standards," Part 43 (1979).
21. Method 2037, Bond Strength, MIL-STD-750B, Test Methods for Semiconductor Devices, Notice 9 (September 19, 1978).
22. Method 2011.3, Bond Strength, MIL-STD-883B, Test Methods and Procedures for Microelectronics, Notice 2 (May 16, 1979).
23. G. G. Harman and C. A. Cannon, The microelectronic wire bond pull test—How to use it, how to abuse it, IEEE Trans. Components, Hybrids Manufacturing Technol. CHMT-1, 203–210 (1978).
24. Thermal Resistance Measurements of Conduction Cooled Power Transistors, EIA Recommended Standard RS-313-B (Revision of RS-313-A) (1975).
25. Method 3131.1, Thermal Resistance, MIL-STD-750B, Test Methods for Semiconductor Devices. Notice 9 (September 19, 1978).
26. D. L. Blackburn, "Semiconductor Measurement Technology," Safe Operating Area Limits for Power Transistors—Videotape Script, NBS Spec. Publ. 400-44. National Bureau of Standards, Washington, D.C., 1977.
27. J. M. Jerke, (ed.), "Semiconductor Measurement Technology," Accurate Linewidth Measurements on Integrated-Circuit Photomasks, NBS Spec. Publ. 400-43. National Bureau of Standards, Washington, D.C., 1980.
28. D. Nyyssonen, Optical Linewidth Measurements on Wafers, Proc. Soc. Photo-Opt. Instrum. Eng. 135, Dev. Semicond. Microlithogr. III pp 115–119 (1978).
29. J. M. Jerke, Semiconductor Measurement Technology: Optical and Dimensional-

Measurement Problems with Photomasking in Microelectronics, NBS Spec. Pub. 400-20. National Bureau of Standards, Washington, D.C., 1976.

30. G. C. Carver, L. W. Linholm, and T. J. Russell, The Use of Mecroelectronics Test Structures to Characterize IC Materials, Processes, and Processing Equipment, *Solid State Technology* (to appear).

31. H. A. Schafft, "Semiconductor Measurement Technology," ARPA/NBS Workshop III, Test Patterns for Integrated Circuits, NBS Spec. Publ. 400-15. National Bureau of Standards, Washington, D.C., 1976.

32. M. G. Buehler, The D-C MOSFET dopant profile method, *J. Electrochem. Soc.* **127**, 701–704 (1980).

33. G. C. Carver and M. G. Buehler, An Analytical Expression for the Evaluation of Leakage Currents in the Integrated Gated-Diode Electrometer (to be published).

34. This volume, Chapter 4.

35. W. S. C. Chang, M. W. Muller, F. J. Rosenbaum, and C. M. Wolf, "Opportunities and Requirements for a National Center for Research on Submicron Structures." Washington Univ. (May 3, 1976).

36. J. N. Zemel and M. S. Chang, "Needs for a National Research and Resource Center in Submicron Structures (East Coast)," Moore School of Electrical Engineering Rep. Univ. of Pennsylvania, Philadelphia, Pennsylvania (May 10, 1976).

37. R. W. Grow, "Needs for a National Research and Resource Center in Submicron Structures." Univ. of Utah (May 26, 1976).

38. H. Smith, *Midwinter Solid-State Res. Conf., Newport Beach, California* (January 1973).

39. R. P. Feynman, There's plenty of room at the bottom, *Am. Phys. Soc. Ann. Meeting, California Tech., Pasadena, California* (1959).

40. M. Lomask, "A Minor Miracle," p. 46, National Science Foundation. U.S. Government Printing Office Stock No. 038-000-00288-1, Washington, D.C.

41. A. Ruoff, *in Proc. NSF Workshop Opportunities Microstructures Sci. Eng. Technol.*, p. 82. Airlie House, Virginia, 1978.

42. Proceedings of the NSF Workshop, (Ref. 41).

43. "The Effect of Government Policy in Innovation, Competition, and Performance: A Case Study of the Semiconductor Industry," pp. 6-29–6-31.

Chapter 8

VLSI, A Major New Tool for Productivity Improvement

PATRICK E. HAGGERTY,
BRUCE R. MAYO,
CHARLES H. PHIPPS

Texas Instruments Incorporated
Dallas, Texas

THE POTENTIAL FOR ELECTRONICS IN MANUFACTURING PRODUCTIVITY

A principal concern at the present time for all who bear responsibility for economic welfare in the United States is the sharp decrease in productivity improvement in recent years. The average annual growth in real output per hour of all persons in private, nonfarm business over the years 1973–1979 was only 0.5% [1]. This is about a quarter of the productivity

301

improvement attained over the entire span of years since World War II. In reaction to that strikingly inadequate performance, the American Productivity Center, in announcing a conference on productivity research held in April 1980, stated: "There is little debate that productivity growth in the USA has slowed and that slowdown affects inflation, unemployment, the balance of payments, our standard of living, and our national and international economic strengths."

Indeed, many couple this sharp decrease in the ability to improve productivity, impending energy shortages, and sharply increased energy costs as inevitably destroying the long-established ability to improve steadily the standard of living enjoyed by the citizens of this country.

These views, however, may well be unduly pessimistic. The invention of the transistor, leading directly to the subsequent invention of the integrated circuit and the development of VLSI, has not just revolutionized electronics, but has created a potential for new tools whose long-term impact on the way people live and work is likely to be enormous. Haggerty, in a 1977 paper [2], expressed it this way:

> I believe that John Bardeen and Walter Brattain and William Shockley, in inventing the transistor, unquestionably made a major contribution to the welfare of man—a contribution with long-term impact that will mean more to man than the invention of the printing press by Gutenberg. I believe that their work has begun a revolution more exciting and more productive in its implications for the future of man than was the industrial revolution signaled and initiated by the invention of the steam engine by Watt.
>
> Thus far, most of the gains proceeding from the industrial revolution have fundamentally multiplied man's muscle and improved his mobility. With the widespread availability of inexpensive and powerful logic and memory, we are, in fact, multiplying man's mind. When one adds the ability to transmit the information processed or the control impulses generated from one end of the world to another, or out into space at 186,000 miles per second, and to display the the information at another location as a veritable reconstitution of reality, surely the ability to multiply, not just our muscle, but our ability to think, to remember, to describe, to imagine, and to create, must imply that the possibilities for continued future gains are boundless. This power to use logic and memory undoubtedly must become even more significant when societies enter, as the United States is beginning to enter, the so-called post-industrial age. In that environment the necessary foodstuffs and products are produced by a limited percentage of the working population. This power may well be the key to

increasing productivity in the service-oriented areas that, thus far, have seemed far less suceptible to the kind of productivity improvements to which we have become accustomed in agriculture and manufacturing.

In that same 1977 paper, Haggerty examines sixteen specific, existing applications of electronics aimed at improving productivity at Texas Instruments and derives from this admittedly limited data a quantitative relationship between the addition of electronics and improved output per worker. This quantitative relationship was dependent on a concept described as an "active element group" (AEG). Originally, an AEG was defined as an active element such as a transistor plus its immediately associated circuitry, including diodes, resistors, capacitors, relays, inductors, coils, connectors, printed circuit boards, etc. In continuing to use this concept of an AEG as the basic building block of electronic circuitry, a logic gate is defined as one AEG and an active memory bit as one AEG, as in Fig. 1.

Figure 2 illustrates the quantitative relationship established between the relative improvement in the output of the work force and AEGs installed per capita (converted from AEGs per worker). By using the correlation between work force output and energy consumption per capita in Fig. 3, the potential for saving energy as the relative output of the work force improved could be correlated with the quantity of AEGs that would have to be installed to achieve such gains.

This relationship, supplemented by a similar but commensurate relationship established for the consumer sector, is then used to suggest that

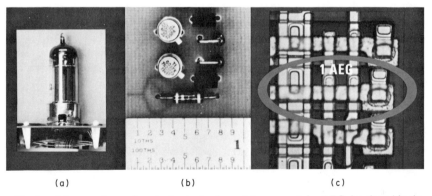

(a)   (b)   (c)

Fig. 1. Advances in semiconductor technology: (a) vacuum tube AEG developed in the mid-1950s (area = 4 in.^{2}); (b) transistor AEG developed in the early 1960s (area = $\frac{3}{4}$ in.^{2}); (c) IC AEG (magnified 1000 times) developed in 1978 (area = $2\frac{1}{2}$ millionths in.^{2}). Copyright Texas Instruments Incorporated.

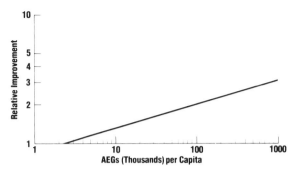

Fig. 2. Plot of AEGs per capita versus relative improvement, showing that adding AEGs improves the output of the work force.

the installation of 1.9×10^{16} AEGs in place in the year 2000 might be so effective in improving productivity as to allow the United States to attain a gross national product of $2611 billion (expressed in constant 1972 dollars) even if energy consumption was limited to early 1970 levels (approximately 70 quads). This is the GNP level projected for the year 2000, if the amount of energy consumed per year increases at 3.7% per year (approximate rate of increase 1966–1976) and, hence, with no depressing effects from energy shortages, as illustrated in Fig. 4.

While Haggerty's 1977 paper does suggest that the new electronics could introduce a technological step function in improving productivity, the correlations established are admittedly speculative and the conclusions drawn simplistic. This chapter attempts to arrive at less simplistic conclusions by narrowing the impact area to the U.S. manufacturing sector and by improving the breadth and quality of the analysis.

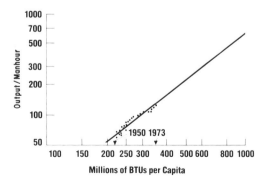

Fig. 3. Correlation of U.S. work force output and energy consumption per capita based on a relative output per manhour in 1967 of 100. The data points cover the period 1950–73.

```
GNP .............................$2,611 Billion (1972 $)

Projected Energy Consumption
    3.7% per Year Rate .......................178 Quads
    Return to Early 1970s Usage ............... 70 Quads
    Reduction .............................108 Quads

AEGs Required
    Cumulative .............................1.9x10¹⁶
    Incremental.............................0.4x10¹⁶
```

Fig. 4. Projection of GNP and related energy consumption in the United States for the year 2000.

Berndt and Wood have developed an economic model that relates the variations in the input factors of labor, capital, energy, and material to manufacturing output [3]. Wood has modified the model to reflect the change in output attributed to technology. The Wood model is discussed in some detail in Appendix I.

As described, the model is structured on the assumption that a reduction in manufacturing output prices can be achieved either by reducing the price of the input factors (capital services K, labor L, energy E, and intermediate materials M), or increasing the efficiency of fewer units of inputs by applying technology.

The model uses 1947–1971 data on manufacturing input factor prices to determine the yearly fractional change in input price necessary to yield the given output prices. This yearly fractional change is attributed to the effects of technology applied to the input factors. These data were also used to derive constants representing the interaction among the various input factors. With this information, output prices are approximately related to input factor prices and a rate of change attributed to technology. Output value and quantity then can be calculated for given demand elasticity. From this data, input quantities for each factor can be found.

The complete model that results after estimating the parameters as described in Appendix I is [cf. Eqs. (13)–(16)]:

$$
\begin{aligned}
\ln U_T = P_Y Y_T / Y_T = {} & 0.0594 \ln P_K + 0.2589 \ln P_L + 0.0434 \ln P_E \\
& + 0.6382 \ln P_M + 0.0191 (\ln P_K)^2 \\
& + 0.0718 (\ln P_L)^2 + 0.0215 (\ln P_E)^2 \\
& + 0.0946 (\ln P_M)^2 + 0.0162 \ln P_K \ln P_L \\
& - 0.0073 \ln P_K \ln P_E - 0.0470 \ln P_K \ln P_M \\
& - 0.0266 \ln P_L \ln P_E - 0.1331 \ln P_L \ln P_M \\
& - 0.0090 \ln P_E \ln P_M + \alpha_T T + \beta_{KT} T \ln P_K \\
& + \beta_{LT} T \ln P_L + \beta_{ET} T \ln P_E + \beta_{MT} T \ln P_M \\
& + \tfrac{1}{2} \beta_{TT} T^2
\end{aligned}
\tag{1}
$$

where $U_T = P_Y$ is the price of unit output, Y_T the quantity of output units, P_K the market price of capital service, P_L the market price of labor, P_E the market price of energy, and P_M the market price of materials. All factors are measured at the end of time span T and α_T, β_{KT}, β_{LT}, β_{ET}, β_{MT}, and β_{TT} are given by [cf. Eqs. (14)–(16)]

$$
\begin{bmatrix} \alpha_T \\ \beta_{KT} \\ \beta_{LT} \\ \beta_{ET} \\ \beta_{MT} \end{bmatrix} =
\begin{bmatrix}
(\alpha_K) & (\alpha_L) & (\alpha_E) & (\alpha_M) \\
0.0594 & 0.2589 & 0.0434 & 0.6382 \\
(\beta_{KK}) & (\beta_{KL}) & (\beta_{KE}) & (\beta_{KM}) \\
0.0381 & 0.0162 & -0.0073 & -0.0470 \\
(\beta_{KL}) & (\beta_{LL}) & (\beta_{LE}) & (\beta_{LM}) \\
0.0162 & 0.1435 & -0.0266 & -0.1331 \\
(\beta_{KE}) & (\beta_{LE}) & (\beta_{EE}) & (\beta_{EM}) \\
-0.0073 & -0.0266 & 0.0429 & -0.0090 \\
(\beta_{KM}) & (\beta_{LM}) & (\beta_{EM}) & (\beta_{MM}) \\
-0.0470 & -0.1331 & -0.0090 & 0.1891
\end{bmatrix}
\times
\begin{bmatrix} \gamma_K \\ \gamma_L \\ \gamma_E \\ \gamma_M \end{bmatrix}, \quad (2)
$$

$$\beta_{TT} = \beta_{KT}\gamma_K + \beta_{LT}\gamma_L + \beta_{ET}\gamma_E + \beta_{MT}\gamma_M, \quad (3)$$

where γ_j, j = K, L, E, M, is the price diminution factor due to technology; $e^{\gamma_j T} - 1$ the annual rate of change in the effective (not market) price of the input factor due to technology enhancements; and $e^{\gamma_j T}$ the total effective input factor price diminution occurring over a span of T years. Then [cf. Eq. (17)]

$$
\begin{bmatrix} S_{K,T} \\ S_{L,T} \\ S_{E,T} \\ S_{M,T} \end{bmatrix} =
\begin{bmatrix}
(\alpha_K) & (\beta_{KK}) & (\beta_{KL}) & (\beta_{KE}) & (\beta_{KM}) & \beta_{KT} \\
0.0594 & 0.0381 & 0.0162 & -0.0073 & -0.0470 \\
(\alpha_L) & (\beta_{KL}) & (\beta_{LL}) & (\beta_{LE}) & (\beta_{LM}) & \beta_{LT} \\
0.2589 & 0.0162 & 0.1435 & -0.0266 & -0.1331 \\
(\alpha_E) & (\beta_{KE}) & (\beta_{LE}) & (\beta_{EE}) & (\beta_{EM}) & \beta_{ET} \\
0.0434 & -0.0073 & -0.0266 & 0.0429 & -0.0090 \\
(\alpha_M) & (\beta_{KM}) & (\beta_{LM}) & (\beta_{EM}) & (\beta_{MM}) & \beta_{MT} \\
0.6382 & -0.0470 & -0.1331 & -0.0090 & 0.1891
\end{bmatrix}
\times
\begin{bmatrix} 1.0 \\ \ln P_K \\ \ln P_L \\ \ln P_E \\ \ln P_M \\ T \end{bmatrix}, \quad (4)
$$

where $S_{j,T}$, j = K, L, E, and M, is the fractional part of the total input value supplied by input factor j at the end of a span of T years. For the input factors [cf. Eqs. (21)],

$$
\begin{aligned}
K_T &= (S_{K,T} \times U_T \times Y_T)/P_K, & L_T &= (S_{L,T} \times U_T \times Y_T)/P_L, \\
E_T &= (S_{E,T} \times U_T \times Y_T)/P_E, & M_T &= (S_{M,T} \times U_T \times Y_T)/P_M,
\end{aligned} \quad (5)
$$

where K_T, L_T, E_T, and M_T are the quantities of the input factors utilized at the end of time span T and [cf. Eq. (20)]

$$Y_T = O_q e^{rT}(U_T/\bar{U}_T)^{-\eta}, \quad (6)$$

where O_q is the output quantity at the beginning of time span T, r the natural growth rate in output quantity, $\bar{U}_T$ the output price at the end of time span T if the rate of price increases producing the natural growth rate persisted until the end of time span T, and η the price demand elasticity.

To solve this model requires estimates of the input prices P_K, P_L, P_E, P_M; factor augmentation rates γ_K, γ_L, γ_E, γ_M; output elasticity η; and natural output growth rate r. The model is used to project the performance of the U.S. manufacturing sector in the year 2000 under a variety of conditions, including the augmentation of labor through the use of AEGs.

For the first set of conditions (Case I), it is assumed that the four input factor prices for capital P_K, labor P_L, energy P_E, and material P_M continue to increase at exactly the same rates as they increased over the model base period, 1947–1971; i.e., P_K, 0.8%; P_L, 4.3%; P_E, 2.1%; and P_M, 1.8%. The output growth rate in the model over the base period is then 3.6%/year, and this same growth rate is assumed for the Case I period, 1970–2000.

These results are summarized in Table I in the column headed Case I, which uses the model to project the manufacturing sector as it would exist in the year 2000 had the conditions prevailing over the base period 1947–1971 prevailed through to the end of the century. Since the 1947–1971 performance of the U.S. economy in general, including the manufacturing sector, can be considered as relatively satisfactory, Case I portrays what might well have been expected of the manufacturing sector in the year 2000 had the energy problem, high inflation rates, and unsatisfactory productivity improvement not occurred during the 1970s.

However, as discussed in Appendix I, the model assumes that because of educational attainment and demographic factors the quality of each labor hour expended improves at 0.62%/year from 1947 through 1971. It would appear, however, that the combination of demographic and educational attainment factors over the span of time from 1970 to 2000 are likely to improve labor quality at only approximately one-quarter the rate of the model's base period. As a consequence, without any of the negative effects of the 1970s being introduced, it is necessary to correct the projection for the year 2000 by changing this labor quality improvement percentage from 0.62%/year to 0.155%/year. The performance of the U.S. manufacturing sector in the year 2000 projected with this single additional negative factor is summarized in Table I under the column heading Case IA.

Of course, during the 1970s the U.S. economy did, in fact, experience the energy price spiral, unhappily high inflation rates, and unsatisfactory productivity improvement. In Case II, the model is used to project the radical changes in the manufacturing sector that these markedly higher rates of price increases of the 1970s would cause. As tabulated under Case

TABLE I

Manufacturing Sector Input and Output Factors for the Year 2000[a]

Factor	Base 1970	Case I	Growth rate (%)	Case IA	Growth rate (%)	Case II	Growth rate (%)	Case III	Growth rate (%)
Output quantity	623.46	1801.4	3.60	1801.4	3.60	1801.4	3.60	1801.4	3.60
Output value[b]	$623.46	$3085.8	5.48	$3218.0	5.62	$19,267.8	12.12	$9993.6	9.69
Output price ($)	1.0	1.713		1.786		10.696		5.548	
Input price factor									
Capital, P_K	1.0	1.27	0.80	1.27	0.80	9.02	7.61	4.59[c]	4.20
Labor, P_L	1.0	3.53	4.30	4.04	4.77	9.60	7.83	6.80[c]	6.07
Energy, P_E	1.0	1.88	2.10	1.88	2.10	33.61	12.43	12.51[c]	7.26
Material, P_M	1.0	1.72	1.80	1.72	1.80	12.39	8.75	6.27[c]	5.28
Input quantity									
Capital, X_K	32.76	106.4		116.6		44.6		62.7	
Labor, X_L	185.5	262.9		255.0		274.1		284.1	
Energy, X_E	27.1	74.4		71.4		63.2		70.4	
Material, X_M	378.1	1094.4		1107.6		1138.5		1100.1	
Shares									
Capital, S_K	0.0526	0.0438		0.0460		0.0209		0.0288	
Labor, S_L	0.2976	0.3008		0.3202		0.1366		0.1932	
Energy, S_E	0.0435	0.0453		0.0417		0.1103		0.0882	
Material, S_M	0.6064	0.6100		0.5920		0.7321		0.6898	

[a] Note: The term η in Eq. (6) has been taken equal 0 in all cases on the assumption that the manufacturing sector rate of price increase is equal to the rate of price increase in the remaining sectors of the economy.

[b] Billions of current dollars.

[c] Growth rates shown for 1980–2000 only. For 1970–1979 period, Case II price factor growth rates are used.

II, the input annual price increases were, over the 1970s, for capital P_K, 7.61%; for labor P_L, 7.83%; for energy P_E, 12.43%; and for material P_M, 8.75%. Thus for the same output quantity (1801.4), the output price increases 10.696 times under the conditions of Case II, as contrasted with 1.713 times under Case I.

Since Case II may well be considered a worst-case projection, the performance of the manufacturing sector also is projected, assuming that the rates of price increases from 1980 to 2000 are halfway between those that occurred during the base period 1947–1971 and the high rates of 1970–1979. Thus Case III uses the input price increases that actually occurred over the 1970–1979 period, as tabulated under Case II for that span of years, and then the ameliorated rates listed under Case III for the remainder of the period, 1980–2000.

Thus Case I can be thought of as projecting a highly desirable performance for the manufacturing sector and Case II a worst-case performance, with Case III projections representing a more likely result, assuming that the high rates of price increases of the 1970s are ameliorated.

Table II summarizes the projections for manufacturing employment under all four cases. Note that under the circumstances of Case I, 22.4 million manufacturing workers would turn out the same manufacturing output quantity as 27.4 million under Case II, and 28.5 million under Case

TABLE II

Projected Manufacturing Employment

Factor	1970	Case I	Case IA	Case II	Case III
Output quantity	623.46	1801.4	1801.4	1801.4	1801.4
Employed workers (millions)	78.6	119.0[b]	119.0	119.0	119.0
Manufacturing workers (millions)[a]	19.4	22.4	25.4	27.4	28.5
Percentage of employed workers	24.7	18.8	21.3	23.0	24.0
Output per worker	32.14	80.42	70.92	65.75	63.21
Productivity improvement in output per worker, 1970–2000 (%)		3.10	2.67	2.41	2.28

[a] Due to physical inventory and other adjustment factors in the model, the model's 1970 labor quantity index relates to a 20.8 million manufacturing work force. The actual 1970 manufacturing work force, 19.4 million was used as the basis for Case I, II, and III projections. Thus, Case I, II, and III manpower projections are somewhat lower than obtained by using the model's base but the ratios are in the right proportion.

[b] Employed work force in 2000 based on: population, 260 million; civilian work force, 127 million; and recent history of 94% employed workers of civilian work force.

III. Again, under the conditions of Case I, the quantity index of output per worker increases from 32.14 in 1970 to 80.42 or 2.5 times, while under Cases II and III it increases to approximately 64, or just over two times. It may seem surprising that the growth rate in output per worker under Case III is slightly less than under Case II and that, consequently, nearly one million more workers would be required to turn out the same output.

Actually, this represents the optimization of input factors introduced by the model on the basis of 1947 to 1971 and reflects the changed relationships of the cost of labor to the other input factors. Thus, under Case III, while the price of labor is increasing at a slower rate than it is under Case II, its relative rate of increase to the other three input price factors is higher. The relative shifts in value of the input factors (price times quantity) must be considered to judge the relative desirability of the various results. These are summarized in Table III for the year 2000. Note particularly that the labor value per index unit of output quantity is 1.461 for Case II and 1.072 for Case III. Because the corresponding labor value per index unit of output quantity for Case I is 0.515, the labor value per index unit of output quantity is somewhat more than doubled for Case III and nearly tripled for Case II.

This deterioration from the base Case I suggests that it would be desirable to examine how the use of AEGs might impact labor productivity to restore the labor value per index unit of quantity under Cases II and III to the lower value of Case I. To achieve comparison, eighteen actual installations of electronics to improve productivity through the use of AEGs have been examined. These applications are described generally in Appendix II, and, of necessity, they are pre-VLSI in complexity and power. They were divided into three categories by type of application, namely,

TABLE III
Factor Values per Output Quantity, Year 2000

Factor	Case I	Case II	Case III
Labor value[a]	928.0	2631.0	1930.35
Labor value/quantity	0.515	1.461	1.072
Energy value ($)	139.9	2124.2	880.4
Energy value/quantity	0.078	1.179	0.489
Material value ($)	1882.4	14.106	6893.2
Material value/quantity	1.045	7.830	3.827
Capital value ($)	135.1	402.1	287.7
Capital value/quantity	0.075	0.223	0.160
Factor value/quantity ($)			
(Output price)	1.713	10.693	5.548

[a] Billions of current dollars.

(1) operators and assemblers, (2) inspectors and crafts, and (3) all other manufacturing workers, including office and professional, and the actual results achieved in these groups of applications also are summarized in Appendix II along with the impact that might be expected if the eighteen applications were expanded to represent the entire manufacturing sector. Although these applications were almost completely aimed at improving the productivity of labor, in the process of achieving that improvement, the productivity of capital also improved significantly and that of energy moderately. These specific applications had little or no impact on improving the productivity of the intermediate materials. In expanding the applications to simulate their impact on the entire manufacturing sector, because they were aimed at improving labor productivity, the potential for improving capital and energy productivity was estimated by assuming 100% penetration of all possible labor applications in the year 2000 and then the impact on capital and energy productivity was scaled to match the mix actually possible in the 18 cases. Because the eighteen cases included a somewhat smaller capital service cost than that for the manufacturing sector as a whole in 1979, the impact of the AEG applications was scaled down accordingly. Similarly, because the energy component in the eighteen cases was markedly lower than that in the manufacturing sector as a whole, it also was scaled down accordingly. As shall be demonstrated in Appendix II, the average impact of the applications for the manufacturing sector as a whole would result in decreasing labor 63.2%, capital 17.2%, energy 5.1%, and materials 0%.

For Cases IV and V, shown in Table IV, it is assumed that AEGs were applied in the manufacturing sector to improve performance on the basis of these relative gains per application, for Cases II and III, respectively.

The impact of the AEGs on labor productivity is shown in Table V. In Case IV, AEGs are applied to Case II to the degree necessary to restore the labor value per index unit of quantity to the level it would have attained under Case I (i.e., a reduction from 1.461 to 0.515). Note that 100% of the potential labor applications are penetrated to achieve this result. Similarly, in Case V, AEGs are applied until Case III's labor value per index unit of quantity is reduced to that of Case I (i.e., from 1.113 to 0.515). To achieve this result, just under 83% of the potential labor applications must be penetrated.

The overall improvement attained with such intensive applications of AEGs is significant. Note that in Case IV the productivity improvement, i.e., the improvement in output per worker, reaches 6.35%/year and in Case V, 4.99%/year.

Clearly then these projections suggest that the potential productivity improvements that could result from intensive application of VLSI are

TABLE IV

Productivity Improvement from AEG Labor Applications to Cases II and III

Factor	1970–1979 Price increases extended to year 2000		1970–1979 Price increases and ameliorated price increases, 1980–2000	
	Case II	Case IV	Case III	Case V
Output quantity	1801.4	1801.4	1801.4	1801.4
Output value[a]	19,267.8	17,446.0	9993.6	8933.6
Output price	$10.696	$9.700	$5.548	4.964
Input price factor				
Capital, P_K	9.02	9.02	4.59	4.59
Labor, P_L	9.60	9.60	6.80	6.80
Energy, P_E	33.61	33.61	12.51	12.51
Material, P_M	12.39	12.39	6.27	6.27
Input quantity				
Capital, X_K	44.6	36.9	62.7	53.8
Labor, X_L	274.1	96.6	284.1	136.5
Energy, X_E	63.2	61.9	70.4	69.2
Material, X_M	1138.5	1138.5	1100.1	1100.1
Shares				
Capital, S_K	0.0209	0.0191	0.0288	0.0276
Labor, S_L	0.1366	0.0532	0.1932	0.1039
Energy, S_E	0.1103	0.1193	0.0882	0.0969
Material, S_M	0.7321	0.8086	0.6898	0.7716

[a] Billions of current dollars.

very large. All the applications are based on pre-VLSI technology and are certainly primitive by comparison with those that will become possible toward the end of the century.

Further, it was unexpectedly difficult to obtain complete enough information on a large number of actual installations because the impact of the AEGs, not just on labor productivity, but on capital and energy as well, was desired. The examples actually selected include assembly operations where the improvement in output per worker approached 20 times and where large numbers of workers were affected. However, for many of the applications selected, the installation, in fact, impacts only a portion of the workers' task (for office or maintenance workers, for example). Thus even though the gain in output for that particular task may be large, the overall increase in output per worker may be modest. It is reasonable to assume, therefore, that the examples should provide a conservative result and represent, on the average, an early and relatively primitive application technology.

TABLE V

Summary Data

Factor	1970–1979 Price inflation extended to year 2000		1970–1979 Price inflation and ameliorated price increases, 1980–2000	
	Case II	Case IV	Case III	Case V
Output quantity	1801.4	1801.4	1801.4	1801.4
Labor value/quantity	1.461	0.515	1.072	0.515
Manufacturing workers (millions)	27.4	8.8	28.5	13.0
Percentage of employed workers	23.0	7.4	24.0	10.9
Output per worker	65.75	204.02	63.21	138.57
Productivity improvement 1970–2000 (%)	2.41	6.35	2.28	4.99
Percentage of AEG penetration of labor applications		100		82.2
AEGs/year in 2000 ($\times 10^{15}$)		660		455
Approximate percentage of AEG produced in year 2000		2.2		1.5

This judgment tends to be confirmed by the data plotted in Fig. 5. Note the marked relative improvement as the number of AEGs per worker goes up for the thirty-seven applications plotted. In approximately one-third of the installations, when 1 to 10 million AEGs per worker were used, productivity gains of 10 to 30 times per worker were attained. When VLSI makes the availability of 100,000 or more AEGs per chip commonplace at a cost of 0.001 to 0.0001 ¢/AEG, both the potential for and likelihood of attaining relative improvements per worker of 10 to 30 or more times should be common. This contrasts with the average improvement of

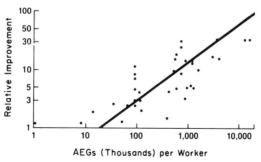

Fig. 5. Plot of AEGs per worker versus relative improvement, showing that adding AEGs improves the output of the work force.

approximately three times for the eighteen applications selected and analyzed.

As VLSI makes available much more powerful and even less costly logic and memory, the power to affect the productivity of labor, capital, energy, and materials will also improve. Furthermore, the eighteen examples analyzed were aimed almost entirely at improving the productivity of labor. Given the very large increases in the price of the other input factors, it would clearly be advantageous to concentrate applications on improving the productivity of capital and especially energy and material. As AEG installations penetrate the potential labor applications successfully, the potential for improving the productivity of capital, energy, and material will deservedly demand higher priorities.

These projections must be considered speculative due to the limitations of the model when input factor prices change so much above those of the base period and due to the very limited information data base that the application cases provide. Almost certainly, however, the general impacts predicted are of the right order of magnitude. The model is based, of course, on the relationships among the four input factors over the base years 1947–1971. Examination of the output quantities projected for the year 2000 suggests that the model probably is being forced past its capabilities to some extent, when the radically different rates of price increases of the 1970s are introduced. It seems unlikely that if the much larger price increases of the 1970s for all four categories are sustained through to the year 2000, the relatively modest amount of capital projected would, in fact, represent an optimum selection. Nevertheless, the model no doubt is approximating the shift among the four factors called for under these circumstances and the relative changes introduced when AEGs are added in Cases IV and V are about right.

Thus the long-range opportunity for improving productivity through the use of VLSI would appear to be very large, and this just emerging technology should have the potential for a significant impact on the lagging productivity that the U.S. is now experiencing.

APPENDIX I. ESTIMATING THE TRANSLOG COST FUNCTION FOR U.S. MANUFACTURING WITH FACTOR AUGMENTATION

A. Aggregate Production and Cost Functions

David O. Wood has drawn upon unpublished work by himself and Ernest R. Berndt to develop a general model of production input factor usage for U.S. manufacturing that generalizes the interaction among inputs and provides a useful treatment of factor augmentation and aggregate technical progress.

In his model, Wood assumes the existence of an aggregate production function that is twice differentiable and characterized by constant returns to scale:

$$Y = F(K^*, L^*, E^*, M^*),$$ (7)

where Y is the gross output, K^* the capital services measured in augmented units, L^* the labor services measured in augmented units, E^* the energy services measured in augmented units, and M^* the services of all other intermediate goods measured in augmented units. Corresponding to Eq. (7) there exists a cost function that reflects the production technology:

$$C = G(Y, P_K^*, P_L^*, P_E^*, P_M^*),$$ (8)

where C is total cost and P_K^*, P_L^*, P_E^*, and P_M^* are the prices of the augmented inputs K^*, L^*, E^*, and M^*, respectively.

Wood also assumes that the cost function, Eq. (8), can be closely approximated by the translog function[1]:

$$
\begin{aligned}
\ln C = {} & \ln \alpha_0 + \ln Y + \alpha_K \ln P_K^* + \alpha_L \ln P_L^* + \alpha_E \ln P_E^* + \alpha_M \ln P_M^* \\
& + \tfrac{1}{2}\beta_{KK}(\ln P_K^*)^2 + \beta_{KL} \ln P_K^* \ln P_L^* + \beta_{KE} \ln P_K^* \ln P_E^* \\
& + \beta_{KM} \ln P_K^* \ln P_M^* + \tfrac{1}{2}\beta_{LL}(\ln P_L^*)^2 + \beta_{LE} \ln P_L^* \ln P_E^* \\
& + \beta_{LM} \ln P_L^* \ln P_M^* + \tfrac{1}{2}\beta_{EE}(\ln P_E^*)^2 + \beta_{EM} \ln P_E^* \ln P_M^* \\
& + \tfrac{1}{2}\beta_{MM}(\ln P_M^*)^2.
\end{aligned}
$$ (9)

Linear homogeneity in prices is necessary if output price is to scale directly with input prices. This imposes the following restrictions on Eq. (9):

$$
\begin{aligned}
\alpha_K + \alpha_L + \alpha_E + \alpha_M &= 1, \\
\beta_{KK} + \beta_{KL} + \beta_{KE} + \beta_{KM} &= 0, \\
\beta_{KL} + \beta_{LL} + \beta_{LE} + \beta_{LM} &= 0, \\
\beta_{KE} + \beta_{LE} + \beta_{EE} + \beta_{EM} &= 0, \\
\beta_{KM} + \beta_{LM} + \beta_{EM} + \beta_{MM} &= 0.
\end{aligned}
$$ (10)

B. Factor Augmentation Due to Technical Change

From the point of view of the production function, it has been traditional to specify certain types of technical change as factor augmenting. Viewed from the vantage of the cost function, such factor-augmenting

---

[1] Note that the first terms in this relation contain the individual input factors and the remaining terms allow for pairwise interaction among the inputs.

technical change corresponds with input-price-diminishing technical change. Let us assume that factor-augmenting technical change occurs at a constant exponential rate

$$X_{jt}^* = X_{jt}e^{\mu_j T} \tag{11}$$

where X_{jt}^* are the services of input j at time t measured in augmented units X_{jt} the services of input j at time t measured in natural units, μ_j the constant exponential rate of augmentation for input j, and $T = t - t_0$, where t_0 is an initial point in time. The corresponding specification for input-price-diminishing technical change is

$$P_{jt}^* = P_{jt}e^{\gamma_j T}, \tag{12}$$

where P_{jt}^* and P_{jt} are the prices associated with X_{jt}^* and X_{jt}, respectively, and γ_j is the constant rate of price diminution for input j.

It is, of course, desirable that $P_{jt}^* X_{jt}^* = P_{jt} X_{jt}$, i.e., that input value be invariant to units of augmentation measurement. For this to hold, it is necessary and sufficient that $\gamma_j = -\mu_j$, i.e., that the rate of factor augmentation is the negative of the rate of input price diminution.

C. Cost Function with Price Augmentation Due to Technical Change

Inserting Eq. (12) into Eq. (8) and utilizing the restrictions in Eq. (10), Wood then rewrites the cost function taking account of technical change

$$
\begin{aligned}
\ln C = {} & \ln \alpha_0 + \ln Y + \alpha_K \ln P_K + \alpha_L \ln P_L + \alpha_E \ln P_E \\
& + \alpha_M \ln P_M + \alpha_T T + \tfrac{1}{2}\beta_{KK}(\ln P_K)^2 + \beta_{KL} \ln P_K \ln P_L \\
& + \beta_{KE} \ln P_K \ln P_E + \beta_{KM} \ln P_K \ln P_M + \beta_{KT}T \ln P_K \\
& + \tfrac{1}{2}\beta_{LL}(\ln P_L)^2 + \beta_{LE} \ln P_L \ln P_E + \beta_{LM} \ln P_L \ln P_M \\
& + \beta_{LT}T \ln P_L \\
& + \tfrac{1}{2}\beta_{EE}(\ln P_E)^2 + \beta_{EM} \ln P_E \ln P_M + \beta_{ET}T \ln P_E \\
& + \tfrac{1}{2}\beta_{MM}(\ln P_M)^2 + \beta_{MT}T \ln P_M + \tfrac{1}{2}\beta_{TT}T^2,
\end{aligned}
\tag{13}
$$

where

$$\alpha_T = \alpha_K \gamma_K + \alpha_L \gamma_L + \alpha_E \gamma_E + \alpha_M \gamma_M, \tag{14}$$

$$
\begin{aligned}
\beta_{KT} &= \beta_{KK}\gamma_K + \beta_{KL}\gamma_L + \beta_{KE}\gamma_E + \beta_{KM}\gamma_M, \\
\beta_{LT} &= \beta_{KL}\gamma_K + \beta_{LL}\gamma_L + \beta_{LE}\gamma_E + \beta_{LM}\gamma_M, \\
\beta_{ET} &= \beta_{KE}\gamma_K + \beta_{LE}\gamma_L + \beta_{EE}\gamma_E + \beta_{EM}\gamma_M, \\
\beta_{MT} &= \beta_{KM}\gamma_K + \beta_{LM}\gamma_L + \beta_{EM}\gamma_E + \beta_{MM}\gamma_M,
\end{aligned}
\tag{15}
$$

and

$$\beta_{TT} = \beta_{KT}\gamma_K + \beta_{LT}\gamma_L + \beta_{ET}\gamma_E + \beta_{MT}\gamma_M. \tag{16}$$

Equations for estimating these various coefficients must now be derived. If input price and output level are fixed, cost-minimizing input demand functions are obtained by logarithmically differentiating Eq. (13).

$$\partial \ln C / \partial \ln P_j = \alpha_j + \sum_k \beta_{jk} \ln P_k + \beta_{jT} T, \qquad j, k = \text{K, L, E, M.}$$

However,

$$\partial \ln C / \partial \ln P_j = P_j \, \partial C / C \, \partial P_j$$

and by Shepard's lemma[2], $\partial C / \partial P_j = X_j$. Thus

$$P_j \, \partial C / C \, \partial P_j = P_j X_j / C$$

and

$$
\begin{aligned}
S_K &= P_K K / C = \alpha_K + \beta_{KK} \ln P_K + \beta_{KL} \ln P_L + \beta_{KE} \ln P_E \\
&\quad + \beta_{KM} \ln P_M + \beta_{KT} T, \\
S_L &= P_L L / C = \alpha_L + \beta_{KL} \ln P_K + \beta_{LL} \ln P_L + \beta_{LE} \ln P_E \\
&\quad + \beta_{LM} \ln P_M + \beta_{LT} T, \\
S_E &= P_E E / C = \alpha_E + \beta_{KE} \ln P_K + \beta_{LE} \ln P_L + \beta_{EE} \ln P_E \\
&\quad + \beta_{EM} \ln P_M + \beta_{ET} T, \\
S_M &= P_M M / C = \alpha_M + \beta_{KM} \ln P_K + \beta_{LM} \ln P_L + \beta_{EM} \ln P_E \\
&\quad + \beta_{MM} \ln P_M + \beta_{MT} T,
\end{aligned}
\tag{17}
$$

where the total cost $C = P_K K + P_L L + P_E E + P_M M$. The S_j are, of course, the cost shares of the inputs in the total cost of producing Y.

Differentiating Eq. (13) with respect to T,

$$
\begin{aligned}
\partial \ln C / \partial T &= \alpha_T + \beta_{KT} \ln P_K + \beta_{LT} \ln P_L + \beta_{ET} \ln P_E \\
&\quad + \beta_{MT} \ln P_M + \beta_{TT} T.
\end{aligned}
\tag{18}
$$

Equation (18) can be rewritten by taking account of the relations of Eqs. (14)–(16).

$$\partial \ln C / \partial T = S_K \gamma_K + S_L \gamma_L + S_E \gamma_E + S_M \gamma_M. \tag{19}$$

Thus the endogenous rate of total cost diminution (output fixed) is equal to a weighted average of the exogenous rates of input price diminution, the weights being endogenous factor shares.

The quantity of output at the end of time span T can be determined from

$$Y_T = O_q e^{rT} (U_T / \bar{U}_T)^{-\eta} \tag{20}$$

where Y_T is the output quantity at the end of time span T, O_q the output

---

[2] Shepard's lemma [7], $\partial C / \partial P_j = X_j$, is the condition that must be satisfied if the Eq. (17) shares are to minimize the output cost.

quantity at the beginning of time span T, r the natural growth rate in output quantity, $\bar{U}_T$ the output price at the end of time span T if the rate of price increases producing the natural growth rate persisted until the end of time span T, U_T the output price at the end of time span T, and η the price demand elasticity.

The quantities of the input factors at the end of time span T can be obtained from

$$K_T = (S_{K,T} \times U_T \times Y_T)/P_K, \qquad L_T = (S_{L,T} \times U_T \times Y_T)/P_L,$$
$$E_T = (S_{E,T} \times U_T \times Y_T)/P_E, \qquad M_T = (S_{M,T} \times U_T \times Y_T)/P_M, \qquad (21)$$

where K_T, L_T, E_T, and M_T are the quantities of the input factors utilized at the end of time span T.

D. Evaluating Historical Constants for the Cost Function

The data used in this study are the same data used in Berndt–Wood [3] and are reproduced in Tables VI and VII. Additional data required are a time series on total factor productivity. Based on the Faucett data [4], Di-

TABLE VI

Price and Quantity Indexes of Capital, Labor, Energy, and Other Intermediate Inputs— U.S. Manufacturing, 1947–1971[a]

Year	P_K	P_L	P_E	P_M	K	L	E	M
1947	1.00000	1.00000	1.00000	1.00000	1.00000	1.00000	1.00000	1.00000
1948	1.00270	1.15457	1.30258	1.05525	1.14103	.97501	.92932	.88570
1949	.74371	1.15584	1.19663	1.06225	1.23938	.92728	1.01990	.94093
1950	.92497	1.23535	1.21442	1.12430	1.28449	.98675	1.08416	1.07629
1951	1.04877	1.33784	1.25179	1.21694	1.32043	1.08125	1.18144	1.13711
1952	.99744	1.37949	1.27919	1.19961	1.40073	1.13403	1.18960	1.17410
1953	1.00653	1.43458	1.27505	1.19044	1.46867	1.20759	1.28618	1.30363
1954	1.08757	1.45362	1.30356	1.20612	1.52688	1.13745	1.29928	1.18144
1955	1.10315	1.51120	1.34277	1.23835	1.58086	1.19963	1.33969	1.32313
1956	.99606	1.58186	1.37154	1.29336	1.62929	1.23703	1.41187	1.35013
1957	1.06321	1.64641	1.38010	1.30703	1.72137	1.23985	1.52474	1.35705
1958	1.15619	1.67389	1.39338	1.32699	1.80623	1.16856	1.44656	1.25396
1959	1.30758	1.73430	1.36756	1.30774	1.82065	1.25130	1.54174	1.41250
1960	1.25413	1.78280	1.38025	1.33946	1.81512	1.26358	1.56828	1.40778
1961	1.26328	1.81977	1.37630	1.34319	1.83730	1.24215	1.59152	1.39735
1962	1.26525	1.88531	1.37689	1.34745	1.84933	1.29944	1.65694	1.48606
1963	1.32294	1.93379	1.34737	1.33143	1.87378	1.32191	1.76280	1.59577
1964	1.32798	2.00998	1.38969	1.35197	1.91216	1.35634	1.76720	1.64985
1965	1.40659	2.05539	1.38635	1.37542	1.98212	1.43460	1.81702	1.79327
1966	1.45100	2.13441	1.40102	1.41878	2.10637	1.53611	1.92525	1.90004
1967	1.38617	2.20616	1.39197	1.42428	2.27814	1.55581	2.03881	1.95160
1968	1.49901	2.33869	1.43388	1.43481	2.41485	1.60330	2.08997	2.08377
1969	1.44957	2.46412	1.46481	1.53356	2.52637	1.64705	2.19889	2.10658
1970	1.32464	2.60532	1.45907	1.54758	2.65571	1.57894	2.39503	2.03230
1971	1.20177	2.76025	1.64689	1.54978	2.74952	1.52852	2.30803	2.18852
Average annual growth rate, 1947–1971	0.8	4.3	2.1	1.8	4.3	1.8	3.5	3.3

[a] Source: Berndt and Wood [3], with permission of North-Holland Publ., Amsterdam.

TABLE VII

Total Cost[a] and Cost Shares of Capital, Labor, Energy, and
Other Intermediate Materials — U.S. Manufacturing, 1947–1971

Year	Total Input Cost[a]	Cost Shares			
		K	L	E	M
1947	182.373	.05107	.24727	.04253	.65913
1948	183.161	.05817	.27716	.05127	.61340
1949	186.533	.04602	.25911	.05075	.64411
1950	221.710	.04991	.24794	.04606	.65609
1951	255.945	.05039	.25487	.04482	.64992
1952	264.699	.04916	.26655	.04460	.63969
1953	291.160	.04728	.26832	.04369	.64071
1954	274.457	.05635	.27167	.04787	.62411
1955	308.908	.05258	.26465	.04517	.63760
1956	328.286	.04604	.26880	.04576	.63940
1957	338.633	.05033	.27184	.04820	.62962
1958	323.318	.06015	.27283	.04836	.61866
1959	358.435	.06185	.27303	.04563	.61948
1960	366.251	.05788	.27738	.04585	.61889
1961	366.162	.05903	.27839	.04640	.61617
1962	390.668	.05578	.28280	.04530	.61613
1963	412.188	.05601	.27968	.04470	.61962
1964	433.768	.05452	.28343	.04392	.61814
1965	474.969	.05467	.27996	.04114	.62423
1966	521.291	.05460	.28363	.04014	.62163
1967	540.941	.05443	.28646	.04074	.61837
1968	585.447	.05758	.28883	.03971	.61388
1969	630.450	.05410	.29031	.03963	.61597
1970	623.466	.05255	.29755	.04348	.60642
1971	658.235	.04675	.28905	.04479	.61940

[a] Source: Berndt and Wood [3], with permission of North-Holland Publ., Amsterdam.

visia indexes of total output (sales of total manufacturing to all sectors of the U.S. economy, including final demand) and Divisia indexes of total factor input (purchases of total manufacturing from all sectors of the U.S. economy, plus labor and capital services) are computed separately. Following Jorgenson and Griliches [5,6], a measure of total factor productivity is obtained as the Divisia index of output minus a Divisia index of inputs. Since the Divisia index is a chained index, one observation is lost. The data for this study are therefore from the period 1948–1971. Table VIII presents the index of the rate of total cost dimunition for U.S. manufacturing, 1948–1971, which is computed simply as the negative of the rate of total factor productivity. If technical progress has taken place, the expected sign of the rate of total cost diminution is negative.

To estimate the parameters of the share Eq. (17), they and Eq. (18) are reparameterized in terms of the constant rates of factor augmentation expressed in Eqs. (14)–(16) and the restrictions summarized in Eq. (10) are imposed. These reparameterized equations are in terms of the β_{ij} and γ_j,

TABLE VIII

Rate of Total Cost Diminution ($\dot{C}/C$) in U.S. Manufacturing, 1948–1971

Year	$\dot{C}/C$	Year	$\dot{C}/C$	Year	$\dot{C}/C$	Year	$\dot{C}/C$
1948	0.01525	1954	0.00555	1960	0.01639	1966	0.00928
1949	0.00500	1955	0.02376	1961	0.00939	1967	0.00577
1950	0.01184	1956	0.01115	1962	0.02443	1968	0.00376
1951	0.01700	1957	0.00782	1963	0.00077	1969	0.01406
1952	0.00029	1958	0.00667	1964	0.03181	1970	0.01241
1953	0.00759	1959	0.00893	1965	0.00815	1971	0.01256

i, j = K, L, E, M. These parameters are estimated using the iterative, nonlinear, minimum distance estimator for simultaneous equations, the procedure employed in the earlier study by Berndt and Wood [3]. The parameter estimates with t-statistics in parentheses are given in Table IX.

It is of considerable interest to examine the estimated rates of factor augmentation (the negative of the estimated γ_j), as 1.05% of K, 1.37% for L, −0.18% for E, and 0.28% for M. The estimated γ_L is significantly different from zero and γ_K is marginally significant. The high estimate for labor is somewhat surprising because the hours-worked data series has already been adjusted for estimated changes in educational attainment over time. Berndt and Wood have assumed that the demographic changes involving the addition of a large number of better-educated, younger people to the work force has resulted in an improvement in the quality of each labor hour worked of 0.62%/year over the years 1947–1971. This has the effect of making each labor hour in 1971 the equivalent of 1.16 labor hours in 1947 and means that 16% fewer workers, in fact, turn out the quantities of equivalent 1947 labor used in generating the model. For the same reason, the price of labor P_L, which the model shows increasing at 4.3%/year, represents an actual increase of 4.92%/year applied to the proportionately smaller number of real labor hours worked.

One intriguing result is that the rate of factor augmentation is largest for labor, the input whose price has risen the most. Thus Berndt and Wood's results suggest that a part of the postwar increase in the demand for energy is because technical change augmented capital, labor, and other intermediate materials more rapidly than it augmented energy. The input energy value developed by the model includes all U.S. domestic and imported crude oil as inputs to petroleum refining. It is not adjusted to only the input energy required for crude oil consumed by the manufacturing sector. Therefore, this input energy value is much larger than either energy consumption or its equivalent energy inputs for the manufacturing sector.

TABLE IX

Parameter Estimates for Translog Model with Factor Augmentation and
Symmetry for U.S. Manufacturing, 1948–1971[a]

	α_j	ij	β_{ij}			
			K	L	E	M
α_K	0.594 (35.3)	K	0.0381 (4.74)			
α_L	0.2589 (35.3)	L	0.0162 (4.74)	0.1435 (2.01)		
α_E	0.0434 (35.3)	E	−0.0073 (1.62)	−0.0266 (1.33)	0.0429 (3.39)	
α_M	0.6382 (35.3)	M	−0.0470 (2.00)	−0.1331 (1.58)	−0.0090 (0.44)	0.1891 (1.77)
α_T	−0.0059 (2.07)					

Augmentation rates		Historical β_{iT} (i = K, L, E, M, T)	
γ_K	−0.0105 (1.91)	β_{KT}	−0.0005 (1.45)
γ_L	−0.0137 (2.84)	β_{LT}	−0.0018 (1.09)
γ_E	+0.0018 (0.30)	β_{ET}	+0.0005 (1.07)
γ_M	−0.0028 (0.89)	β_{MT}	+0.0018 (0.98)
		β_{TT}	0.0+ (0.91)

[a] Values in parentheses are parameter estimates based on t-statistics.

Before leaving the topic of technical change, it is necessary to note that when technical change is assumed to be Hicks-neutral,[3] the estimate of the common rate of factor augmentation is 0.60%, which, as an annual rate of technical change, may at first glance appear a bit low. Spencer Star [8] has shown that previous studies measuring technical change have tended to produce somewhat large figures because they look only at value added (capital and labor) and ignore intermediate materials. In his study based on 1950 and 1960 census data for seventeen U.S. manufacturing industries at the two-digit level, Star found that the overall annual rate of

---

[3] That is, all input factors are augmented at the same rate.

technical change was reduced from 1.51 to 0.59% when, among other things, intermediate inputs were properly taken into account. Incidentally, Star's figure of 0.59% is remarkably close to Berndt and Wood's estimated 0.60%.

APPENDIX II. IMPROVING PRODUCTIVITY THROUGH THE USE OF AEG—EIGHTEEN APPLICATIONS AND THEIR RELATIONSHIP TO U.S. MANUFACTURING

A. Description of the Applications

1. Four Applications Impacting Operator/Assembly Functions

a. Steel Soaking Pit. Steel ingots are soaked in a heated enclosure in order to bring them to a uniform temperature before rolling. Fuel flow, inlet air, and exhaust air are controlled to set pit temperature, pressure, and combustion products. An electronic, programmable controller replaces the conventional hydraulic or pneumatic controller. The flexible optimization possible due to the extensive logic and memory in the programmable controller reduces fuel costs, reduces scaling, increases throughput, and lengthens pit life.

b. Automatic Assembly Machine. This application initially used an assembly machine totally directed by an operator for the attachment of small piece parts. This was replaced with an assembly machine with extensive memory for storing the exact work to be done, a camera for automatically aligning the piece parts to the work station, and logic for making the alignment and assembly decisions. Benefits included greatly increased throughput with the attendant reduction in labor and floor space. Yields rose about 1%.

c. Chemical Mixing. This application compares an old manually controlled plant for mixing chemicals to a new electronically controlled plant. Since the basic plumbings of the old and new plants were similar, all the improvement was attributed to the installation of the electronic controls. In the electronically controlled plant, a minicomputer controls ten weighing tanks, each fed by up to nine materials. The minicomputer computes and controls the quantities of mix required to meet the final solution weight for a given order. Savings are primarily due to decreased labor and increased throughput.

d. Machine Control for Printed Wiring Board Fabrication. Computer control of machine tools was introduced into a printed wiring board fabrication shop. The output per operator increased 35%.

2. Four Applications Impacting Inspection/Crafts

a. Mechanical Subsystem Test. A complex mechanical subsystem with many modes of operation was to be tested. Formerly this was done by an operator manipulating the subsystem and recording the data. Minicomputers were used to automate the running of the tests and the recording of the data. A central host computer made the data available to management and to repair areas. The amount of labor required was reduced and the work load was significantly lightened on the labor that remained. Increased throughput greatly decreased the number of test stands required.

b. Modem Test System. Modems were formerly tested by an operator connecting the proper input signals to the modem and measuring the modem output. A minicomputer was used to automate the test by controlling a switching matrix for routing the desired input signals to the modem under test. Error reports were prepared automatically by the automated test system. Throughput greatly increased and the technical level of the test personnel could be reduced from technician to test operator.

c. Vision-Aided Material Sorter. Material was sorted into various bins depending on the shade of color. A minicomputer–camera system with an associated material-handling system was used to replace a manual sort system. The benefit was a large reduction in the required labor and factory floor space.

d. Computer Test System. Special hardware and diagnostic software were developed for a minicomputer-based system designed to reduce checkout and fault isolation time in unit tests for the board set of a large minicomputer. This reduced the number of test stations by a factor of ten compared to the previous more manual test stations. It also permitted lower job grade technicians to perform the tests.

3. Ten Applications Impacting Other Parts of Manufacturing, Such as Engineering, Planning, Finance, Control, Etc.

a. Microfilm Retrieval System. A high-volume manufacturer of small parts enters sales orders on microfilm. Previously it was a manual search and view to retrieve a desired document. This approach was replaced by a minicomputer-based system that retrieves the proper roll and locates the document on the roll based on a unique number assigned to each document. Benefits primarily were reduced labor.

b. Speaker Verification Access Control. Originally, high-quality guards were used to control access to sensitive areas of a plant. Guards were replaced by a minicomputer system that compared the voice of the person desiring access to stored voice prints. One minicomputer system could control three access points while requiring only a single person to operate and monitor it.

c. Software Development System for Programmers. Initially a software development organization operated in the batch mode. Programmers would write their programs and submit them for running to a central computer dedicated to software development. In the improved system the same central computer was used but a remote computer terminal was placed on each programmer's desk. In this application the productivity change in the programmers as they switched from the batch mode to the real-time mode is measured.

d. Computer System for Recording Inventory and Scheduling Repair Parts at a Service Center. In a calculator repair facility, all record-keeping, tracking, and report generating were initially done manually. A computerized, automated system was installed that utilized a bar code tag with owner information on each returned calculator. The calculator then was tracked through the repair depot by the computer. The computer also prepared invoices, shipping labels, COD cards, insurance logs, and work in progress reports.

e. Field Service Diagnostic and Scheduling System. Thirty-four minicomputers with associated terminals and storage disks were used to provide information to minicomputer field service offices as well as to serve other management functions. It is estimated that 40% of the system was devoted to providing information to field service. Service calls per day per serviceman increased 45%.

f. Interactive Graphics for Integrated Circuit Design. Manual methods of designing integrated circuits were replaced by a computerized system with real-time designer interaction with a vector graphics display. Design rules and aids were incorporated into the software. Greatly decreased design time and reduced errors were achieved.

g. Accounting System for Monitoring Accounts Receivable. Initially, accountants made inputs to a receivable accounting system by batch submissions to a central computer. Then a remote terminal was placed on each accountant's desk so that input/output could be from each desk. This application compares the productivity of the accountants when operating in the remote terminal mode versus the batch mode.

h. Desk Programmable Calculator in Financial and Planning Functions. Persons in financial and planning organizations initially had the option of calculating manually with simple four-function calculators or on large, central computers. To fill the gap between simple four-function calculators and large, central computers, each person was issued a desk-top, programmable calculator with prompting and various software support modules. This application compares the productivity of the financial and planning personnel before and after the issue of the programmable desk-top calculators.

i. Production Control System for Electronic Equipment Assembly. A computerized system was installed on a digital electronics factory floor to control the flow of work and give work in progress status. The ready availability of accurate information increased the output per person by 16%.

j. Material Control System. The improved system uses carousels, a converter–diverter, and a minicomputer system to store, deliver, and track material used in the cable assembly area of a computer manufacturer. The computerized control system provides real-time tracking and report generation. Before installation of the computerize system, material was delivered by hand and tracked by hand-generated lists.

B. Analysis of the Applications' Data

1. Impact on Capital, Labor, and Energy

Data has been accumulated on the eighteen manufacturing sector, electronic productivity examples. For each one, the annual capital, labor, and energy costs were collected for before and after conditions. Costs are expressed in 1979 dollars and are normalized for constant output from the operating entity. These examples were divided into three groups: four projects supported operator and assembly workers; four projects supported inspection and craft workers; and ten projects supported other manufacturing areas such as sales, engineering, planning, finance, and control. The total before and after capital, labor, and energy costs for the three work areas as given by the applications is shown in Table X.

TABLE X
Summary of Eighteen Electronic Automation Applications[a]

Type of worker	Capital services		Labor		Energy	
	Before	After	Before	After	Before	After
Operator/assembly	974.2	632.4	12,678.8	761.6	256.5	150.9
Inspection/craft	2214.3	1032.8	6,120.6	1230.3	19.46	5.26
Other	400.0	518.8	12,774.3	7637.8	58.3	38.32

[a] Thousands of 1979 dollars.

2. Limitations of the Data Sample

The examples that produced the data in Table X do not fully represent the characteristics of the electronic productivity applications that will occur in the next twenty to thirty years in the U.S. manufacturing sector in at least two important ways. First, they were overwhelmingly directed at improvement in labor productivity with the gains in capital and energy productivity as incidental side benefits. Over the next twenty years, a significant portion of new electronic applications will have as their primary purpose improving the productivity of capital, energy, or materials. Second, about 90% of the applications were in light manufacturing. This portion of the manufacturing sector is a relative low user of energy. Thus, while Table X shows a large reduction in energy usage, the amount of energy involved is small.

3. Potential Impact on Total U.S. Manufacturing

a. **Characteristics of U.S. Manufacturing Sector Labor in 1979.** In spite of these limitations, it is worthwhile to investigate what the impact would be on U.S. manufacturing if electronic productivity improvements of the type covered in Table X were to penetrate the entire U.S. manufacturing sector. For 1979 the total number of wage and salary workers in U.S. manufacturing was 20,979,000 [1, p. 242, Table B-34]. Wage and salary disbursement to the manufacturing sector for 1979 was $330.9 billion [1, p. 226, Table B-20]. Since the labor costs in Table X include wages and benefits, the $330.9 billion is increased by 18.9% to $393.4 billion. [1, p. 224, Table B-19]. It is estimated that 40% of the manufacturing workers are in the operator/assembly category, 16% in the inspection/crafts category, and 44% in other manufacturing categories [9]. From the data base that produced Table X, the annual wages plus benefits were calculated for each of the three category of workers. These wages plus benefits were then reduced 3.3% for each category so that the sum of the wages paid for

TABLE XI

Summary of Wages Plus Benefits for the U.S. Manufacturing Sector, 1979

Type of worker	Number of workers (millions)	Annual wages plus benefits per worker ($)	Annual total wages plus benefits (billions of dollars)
Operator/assembly	8.39	14,038	117.7
Inspection/craft	3.36	20,200	67.9
Other	9.23	22,521	207.8

all the workers in the U.S. manufacturing sector would total the previously determined figure of $393.4 billion. The number of workers, wages plus benefits per worker, and the total wages and benefits for the U.S. manufacturing sector are summarized in Table XI.

b. Scaling the Applications Data to the Full U.S. Manufacturing Sector. Scaling up Table X so that the before labor dollars for each worker category is equal to the annual total wages and benefits for that worker category in Table XI produces Table XII, which is the before and after capital, labor, and energy values that would result if electronic productivity applications similar to the examples described in the data of Table X were replicated throughout the U.S. manufacturing sector.

In determining the cost of capital services for Table XII, a 25-year building life, a 7.14-year equipment life, and a 10% interest rate were assumed for all the application projects. This was thought consistent with future trends toward faster depreciation and higher interest rates.

TABLE XII

Data for 100% Penetration of Electronic Productivity Applications of the Type Described by Table I—U.S. Manufacturing Sector, 1979[a]

Type of worker	Capital services		Labor		Energy	
	Before	After	Before	After	Before	After
Operator/assembly	9.04	5.87	117.70	7.07	2.38	1.40
Inspection/craft	24.57	11.46	67.90	13.65	0.22	0.06
Other	6.51	8.44	207.80	124.24	0.95	0.62
Total	40.12	25.77	393.40	144.96	3.55	2.08
Percentage reduction		35.8		63.2		41.4
Percentage of U.S. manufacturing	48.1		100		5.07	

[a] Billions of 1979 dollars.

No figures were readily available for the cost of capital services for U.S. manufacturing for 1979. For 1978, total depreciation for manufacturing buildings and equipment was $40 billion [1, p. 558, Table 922]. The model of the manufacturing sector described in Appendix I predicted a total cost of capital services for 1979 of $83.4 billion. Since the $83.4 billion 1979 cost of capital services seems consistent with the $40 billion depreciation in 1978, it was assumed that the cost of capital services for U.S. manufacturing in 1979 was $83.4 billion, even though this seemed to imply a longer building and equipment life and a lower interest rate than was assumed in Table XII.

According to one source [10], the U.S. manufacturing sector consumes 85% of the U.S. industrial sector energy. Applying this factor to the industrial sector 1979 energy prices and consumption given by the DoE [11] gives approximately $70 billion for the U.S. manufacturing energy costs in 1979.

Using the derived values for total U.S. manufacturing cost of capital services, labor, and energy for 1979, the amount of these costs that is impacted by 100% penetration of electronic automation of the type shown in Table X is given in Table XII.

ACKNOWLEDGMENT

The authors would like to acknowledge the contributions of the work of Dr. David O. Wood to this chapter. The manufacturing sector model equations in the first section and almost all of Appendix I were taken from an unpublished paper by Wood.

REFERENCES

1. "Economic Report of the President," p. 246, Table B-37. U.S. Government Printing Office, Washington, D.C., 1980.
2. P. E. Haggerty, Shadow and Substance. Texas Instruments, Dallas, Texas, 1977.
3. E. R. Berndt and D. O. Wood, Technology, prices and the derived demand for energy, *Rev. Econ. Statis.* **57,** 259–267 (1975).
4. Jack Faucett Associates, Data Development for the I–O Energy Model: Final Report. Jack Faucett Associates, Inc., Chevy Chase, Maryland (1973).
5. D. W. Jorgenson and Z. Griliches, The explanation of productivity change, *Rev. Econ. Stud.* **34,** 249–283 (1967).
6. D. W. Jorgenson and Z. Griliches, Issues in growth accounting: A reply to Edward F. Denison, *Surv. Current Business* **52,** 65–94 (1972).
7. R. Shepard, "The Theory of Cost and Production Functions." Princeton Univ. Press, Princeton, New Jersey, 1970.
8. S. Star, Accounting for the growth of output, *Am. Econ. Rev.* **64,** 123–135 (1974).

9. Statistical Abstracts of the United States 1979, p. 415, Table No. 685. U.S. Department of Commerce.

10. "Energy Alternative—A Comprehensive Analysis," The Science and Public Policy Program, Univ. of Oklahoma, Norman, Oklahoma, Stock No. 041-011-00025-4, pp. 13–22. U.S. Government Printing Office, Washington, D.C., 1975.

11. Monthly Energy Review—July, U.S. Department of Energy, Energy Information Administration, DOE/EIA 0035/80(07), pp. 17–23, 72–85 (1980).

Index